The Paris Framework for Climate Change Capacity Building

The Paris Framework for Climate Change Capacity Building pioneers a new era of climate change governance, performing the foundational job of clarifying what is meant by the often ad-hoc, one-off, uncoordinated, ineffective and unsustainable practices of the past decade described as 'capacity building' to address climate change. As an alternative, this book presents a framework on how to build effective and sustainable capacity systems to meaningfully tackle this long-term problem. Such a reframing of capacity building itself requires means of implementation. The authors combine their decades-long experiences in climate negotiations, developing climate solutions, climate activism and peer-reviewed research to chart a realistic roadmap for the implementation of this alternative framework for capacity building. As a result, this book convincingly makes the case that universities, as the highest and sustainable seats of learning and research in the developing countries, should be the central hub of capacity building there.

This book will be a valuable resource for students, researchers and policymakers in the areas of climate change and environmental studies.

Mizan R. Khan is Professor of Environmental Management at North South University, Bangladesh, and a Lead Author of the Intergovernmental Panel on Climate Change (IPCC) assessment reports. He serves as a lead member in the Bangladesh delegation to the UNFCCC negotiations since 2001.

J. Timmons Roberts is Ittleson Professor of Environmental Studies and Sociology at Brown University, Providence, RI, USA, and founder and leader of the Climate and Development Lab at the Institute at Brown for Environment and Society.

Saleemul Huq is the Director of the International Centre for Climate Change and Development (ICCCAD) in Dhaka, Bangladesh, and a frequent advisor for the Least Developed Countries negotiating group. He has also been a Lead Author of the Intergovernmental Panel on Climate Change (IPCC) assessment reports.

Victoria Hoffmeister is an analyst at Redstone Strategy Group, working as a consultant for climate- and conservation-focused foundations and nonprofits. She worked for three years as a member of the Climate and Development Lab at Brown University, as well as in the White House Council on Environmental Quality during the Obama administration.

Routledge Advances in Climate Change Research

Responding to Climate Change in Asian Cities
Governance for a More Resilient Urban Future
Edited by Diane Archer, Sarah Colenbrander and David Dodman

Climate Change Finance and International Law
Alexander Zahar

Urbanization and Climate Co-Benefits
Implementation of Win-Win Interventions in Cities
Edited by Christopher Hideo Doll and Jose Puppim de Oliveira

Climate Change and Food Security
Africa and the Caribbean
Edited by Elizabeth Thomas-Hope

Globalizing the Climate
COP21 and the Climatisation of Global Debates
Edited by Stefan C. Aykut, Jean Foyer and Edouard Morena

Adapting Infrastructure to Climate Change
Advancing Decision-Making Under Conditions of Uncertainty
Todd Schenk

Collaborative Local Climate Change Action
Pathways and Progress towards Transformation
Edited by Susie Moloney, Hartmut Fuenfgeld and Mikael Granberg

Pricing Carbon in Australia
Contestation, the State and Market Failure
Rebecca Pearse

The Paris Framework for Climate Change Capacity Building
Mizan R. Khan, J. Timmons Roberts, Saleemul Huq and Victoria Hoffmeister

The Paris Framework for Climate Change Capacity Building

Mizan R. Khan, J. Timmons Roberts,
Saleemul Huq and Victoria Hoffmeister

Routledge
Taylor & Francis Group
LONDON AND NEW YORK

earthscan
from Routledge

First published 2018
by Routledge
2 Park Square, Milton Park, Abingdon, Oxon OX14 4RN

and by Routledge
711 Third Avenue, New York, NY 10017

Routledge is an imprint of the Taylor & Francis Group, an informa business

British Library Cataloguing in Publication Data
A catalogue record for this book is available from the British Library

Library of Congress Cataloging in Publication Data
A catalog record for this book has been requested

ISBN: 978-1-138-89664-2 (hbk)
ISBN: 978-1-315-17909-4 (ebk)

Typeset in Sabon
by Wearset Ltd, Boldon, Tyne and Wear

Contents

Illustrations

Tables

Boxes

Contributors

Harry August, Climate and Development Lab, Brown University, USA.

Alex Barba, Climate and Development Lab, Brown University, USA.

Julianna Bradley, Climate and Development Lab, Brown University, USA.

Mara Dolan, Climate and Development Lab, Brown University, USA.

Joseph Epitu, University of Makerere, Uganda.

Caroline Jones, Climate and Development Lab, Brown University, USA.

Jessica Kenny, Climate and Development Lab, Brown University, USA.

Shaila Mahmud, Research Officer at the International Centre of Climate Change and Development (ICCCAD), Dhaka, Bangladesh.

Frishta Qaderi, Climate and Development Lab, Brown University, USA.

Stacy-ann Robinson, Postdoctoral Researcher, Climate and Development Lab, Brown University, USA.

Kai Salem, Climate and Development Lab, Brown University, USA.

Austen Sharpe, Climate and Development Lab, Brown University, USA.

Revocatus Twinomuhangi, University of Makerere, Uganda.

Andrea Zhu, Climate and Development Lab, Brown University, USA.

Aaron Ziemer, Climate and Development Lab, Brown University, USA.

Preface

The 2015 Paris Agreement was widely hailed as a breakthrough in global efforts to address climate change. Little known and barely discussed, however, is Article 11, under which the nations of the world decided to establish the Paris Committee on Capacity Building, and definitively place capacity building as a crucial means to enhance climate action in developing countries. Essentially, the attempt seeks to avoid the way climate and development planning has often been done in the past, which has usually involved experts and consultants from developed countries 'parachuted in', drafting up plans for projects and leaving little room for the developing countries to address their own issues. In addressing the short-term goals of developing projects, little real capacity was left behind.

In hindsight, the initiatives on capacity building led by the United Nations Framework Convention on Climate Change (UNFCCC) began with COP7 in 2001. In response, millions of dollars have been spent in developing countries by bilateral and multilateral development agencies, including the UNFCCC Secretariat, the World Bank-managed Global Environment Facility, the US Agency for International Development (USAID) and the UK Department for International Development (DFID). Millions more are pledged for the coming years.

What has been the end result? What do we mean by capacity building and have these efforts made any progress? How much and what capacities have been built? Who led the process? Were capacity-building efforts demand- or supply-driven? Has there been any sustainable system left in place? If not, what are the gaps and lacunae in the process? And how can they be implemented?

The Paris Framework for Climate Change Capacity Building aims at unearthing the often ineffective and unsustainable practices of capacity building in the last decade to address climate change. As an alternative, the book presents a roadmap on how to build effective and sustainable capacity systems to meaningfully tackle the problem.

The importance and salience of such a book emerged from Article 11 of the Paris Agreement adopted at COP21 of the UNFCCC in December 2015, which establishes the Paris Committee on Capacity Building (PCCB) and stipulates that: 'Capacity building should be guided by lessons learned, including those from capacity building activities under the Convention, and should be an effective, iterative process that is participatory, cross-cutting and gender-sensitive.'

There have been reports by bilateral and multilateral agencies on their capacity building activities, including the UNFCCC synthesis reports, but there is not a single book in the field dedicated to analysing the end result of past capacity building in addressing climate change.

The main theme of the book is learning from the experiences of the past decades of capacity building activities of bilateral and multilateral agencies in developing countries, revealing the gaps and lacunae in those processes, and proposing a new way forward. Our observation is that activities in this area so far have often been disparate and uncoordinated, one-off, short-term consultancy-based initiatives, supported by different agencies, which have spent millions of dollars, but which often failed to build and leave a sustainable long-term system in place. The book makes the case that universities in developing countries should be the central hub of such activities, since they are long-term institutions that will outlast short-term funding cycles and project termination.

Acknowledgements

This book is the result of a major collaboration across continents and universities, and the result of some heroic efforts by our partners. The authors want first to thank the case study researchers and authors Revocatus Twinomuhangi and Joseph Epitu for the Uganda case at the University of Makerere, Stacy-ann Robinson for the Jamaica case, and Shaila Mahmud for assisting with the Bangladesh case. Research and writing for Chapter 6 were led by Julianna Bradley at Brown University, USA, and the work of assembling the data sets for six agencies was done by Climate and Development Lab 2015–16 members Andrea Zhu, Harry August, Austen Sharpe, Caroline Jones, Kai Salem and especially Jessica Kenny. We appreciate the thoughtful comments of our editors and reviewers. Substantial editing and formatting work was done by Julianna Bradley and Sonya Gurwitt, to whom we are extremely grateful. Several CDL 2017–18 members stepped up in the final hours to push the project over the finish line: Frishta Qaderi, Alex Barba, Aaron Ziemer, Mara Dolan, and Harry August. Heartfelt thanks to all.

Abbreviations

ACCC	Adaptation to Climate Change in the Caribbean
ACCRA	Africa Climate Change Resilience Alliance
ADB	Asian Development Bank
AF	Adaptation Fund
AfDB	African Development Bank
AIMS	Aid Information Management System
AR5	Fifth Assessment Report of the IPCC
Art.	Article
ATLAS	Adaptation Assessments, Thought Leadership, and Learning
B2B	Business to Business
BCCRF	Bangladesh Climate Change Resilience Fund
BCCSAP	Bangladesh Climate Change Strategy and Action Plan
BCCTF	Bangladesh Climate Change Trust Fund
BINGOs	business and industry NGOs
BMD	Bangladesh Meteorological Department
BoT	Board of Trustees
BPATC	Bangladesh Public Administration Training Centre
BR	Biennial Report
BUET	Bangladesh University of Engineering and Technology
BURS	Biennial Update Reports
CAES	College of Agricultural and Environmental Sciences
CALIP	Climate Adaptation and Livelihood Protection
CAN-U	Climate Action Network Uganda
CARICOM	Caribbean Community
CBD	Convention on Biological Diversity
CBDR&RC	Common but Differentiated Responsibilities and Respective Capacities
CBIT	Capacity Building Initiative for Transparency
CCAKB	Climate Change Adaptation Knowledge Base
CCAP	Climate Change Action Plan
CCB	Committee on Capacity Building
CCC	Climate Change Centre

CCC	Climate Change Cell
CCD	Climate Change Department
CCIC	Climate Change Information Clearinghouse
CCRD	Climate Change Resilient Development
CCU	Climate Change Unit
CDKN	Climate and Development Knowledge Network
CDM	Clean Development Mechanism
CEDAT	College of Engineering, Design, Art and Technology
CFTM	Climate Finance Transparency Mechanism
CIDA	Canadian International Development Agency
CIF	Climate Investment Funds
CIP	Country Investment Plan
CMA1	First Conference of the Parties serving as the meeting of Parties to the Paris Agreement
COCIS	College of Computing and Information Sciences
COP	Conference of the Parties
COP21	21st Conference of the Parties
CPACCP	Caribbean Planning for Adaptation on Climate Change Project
CPP	Cyclone Preparedness Programme
CREEC	Centre for Research in Energy and Energy Conservation
CSG	Climate Studies Group
CTCN	Climate Technology Centre and Network
CTD	Committee on Trade and Development
CUAS	Cologne University of Applied Sciences
DDAGTF	Doha Development Agenda Global Trust Fund
DDPs	District Development Plans
DFID	Department for International Development
DIE	Deutsches Institut für Entwicklungspolitik
DMCs	developing country members
DNA	Designated National Authority
DoE	Department of Environment
DPM	Disaster Preparedness and Management
DRR	disaster risk reduction
EACREEE	East African Centre for Renewable Energy and Energy Efficiency
EC	Executive Committee
EIF	Enhanced Integrated Framework
ENGOs	environmental non-governmental organizations
ERICCA	Education and Research to Improve Climate Change Adaptation
ERTs	Expert Review Teams
FAO	Food and Agriculture Organization
FSV	Facilitative Sharing of Views
GAR15	Global Assessment Report 2015

GATT	General Agreement of Tariff and Trade
GBM	Ganges/Padma, Brahmaputra/Jamuna, and Meghna Rivers
GCCA	Global Climate Change Alliance
GCF	Green Climate Fund
GCM	General Circulation Model
GED	General Economic Division
GEF	Global Environment Facility
GHG	greenhouse gas
GIZ	German Federal Enterprise for International Cooperation
GNP	gross national product
GOB	Government of Bangladesh
GPB	global public bad
GPG	global public good
HCL	Hydrology and Climate Change
HELCOM	Helsinki Commission
HELIX	High End Climate Impact and Extremes
HFA	Hyogo Framework of Action
HILIP	Haor Infrastructure and Livelihood Improvement Project
HiMAP	High Mountains Adaptation Partnership
HRD	Human resource development
HRTT	Human Rights Task Team
IAR	International Assessment and Review
ICA	International Consultation and Analysis
ICCAD	International Centre for Climate Change and Development
IDA	International Development Association
IDB	Inter-American Development Bank
IF	integrated framework
IFC	International Finance Corporation
IG	intergovernmental organization
IITA	International Institute of Tropical Agriculture
INC	Initial National Communication
INDCs	Intended Nationally Determined Contributions
IPCG	Intergovernmental Panel on Climate Change
IPOs	indigenous peoples organizations
ITTC	Institute for Training and Technical Cooperation
IUB	Independent University of Bangladesh
IWFM	Institute of Water and Flood Management
JITAP	Joint Integrated Technical Assistance Programme
LDCF	Least Developed Countries Fund
LDCs	least developed countries
LUCCC	LDC Universities Consortium on Climate Change
MA	Multilateral Assessment
MACC	Mainstreaming Adaptation to Climate
MAK	Makerere University
MDAs	Ministries, Departments and Agencies

MDBs	multilateral development banks
MDG	Millennium Development Goals
MEAs	multilateral environmental agreements
MF	Multilateral Fund
MoDMR	Ministry of Disaster Management and Relief
MoEF	Ministry of Environment and Forests
MoFPED	Ministry of Finance, Planning and Economic Development
MoLG	Ministry of Local Governments
MOOCs	massive open online courses
MOWCA	Ministry of Women and Children Affairs
MRV	Measurement, Reporting and Verification
MTs	multilateral trade system
MUCCA	Makerere University Climate Change Association
MUCCRI	Makerere University Centre for Climate Research and Innovations
NAPAs	National Adaptation Programmes of Action
NARO	National Agricultural Research Organization
NCCAC	National Climate Change Advisory Committee
NCCP	National Climate Change Policy of 2015
NCCPC	National Climate Change Policy Committee
NCSA	National Capacity Self-Assessment
NDC	Nationally Determined Contribution
NDP	National Development Plan
NEPAD	New Partnership for African Development
NGO	non-governmental organization
NHRAP	national human rights action plan
NIEs	national implementing entities
NORAD	Norwegian Agency for Development Cooperation
NOUs	National Ozone Units
NPA	National Planning Authority
NSDP	National Summary Data Page
NTF	Nordic Trust Fund
ODA	Official Development Assistance
ODPEM	Office of Disaster Preparedness and Emergency Management
ODS	ozone depleting substances
OECD	Organization for Economic Cooperation and Development
OHCHR	Office of the High Commissioner for Human Rights
OPM	Office of the Prime Minister
PCCB	Paris Committee on Capacity Building
PIF	Project Identification Forms
PIOJ	Planning Institute of Jamaica
PLS	Progressive Learning Strategy
PMF	Performance Measurement Framework
PPCR	Pilot Program for Climate Resilience

PV	photovoltaic
PVCs	particularly vulnerable countries
RAN	Resilient Africa Network
RAPs	regional action plans
RCM	Regional Climate Model
REDD	Reducing Emission from Deforestation and Forest Degradation
RINGOs	research and independent NGOs
RPIU	Regional Project Implementation Unit
RSP	Regional Seas Programme
RUFORUM	Regional Universities Forum for Capacity Building in Agriculture
SAP/BIO	Strategic Action Plan for the Conservation of Biological Diversity
SBI	Subsidiary Body for Implementation
SBSTA	Subsidiary Body for Scientific and Technological Advice
SCCF	Special Climate Change Fund
SDC	Swiss Development Cooperation
SDGs	Sustainable Development Goals
SESM	School of Environmental Science and Management
SFDRR	Sendai Framework for DRR
SIDA	Swedish International Development Cooperation Agency
SIDS	Small Island Developing States
SNC	Second National Communication
SPARRSO	The Space and Remote Sensing Agency
SPS	sanitary and phytosanitary
SSTs	sea surface temperatures
STDF	Standards and Trade Development Facility
TBTs	technical barriers to trade
TC	technical cooperation
TFAF	Trade Facilitation Agreement Facility
TNC	Third National Communication
TNGOs	trade unions non-governmental organizations
TRTA	trade-related technical assistance
UAL	Urban Action Innovations Lab
UNCC	Universities Network for Climate Capacity
UNCCD	UN Convention to Combat Desertification
UNDP	United Nations Development Programme
UNEP	United Nations Environment Programme
UNFCCC	United Nations Framework Convention on Climate Change
UNISDR	UN International Strategy for Disaster Reduction
UNMA	Uganda National Meteorology Authority
UNSSC	UN System Staff College
USAID	United States Agency for International Development

UWI	University of the West Indies
VEGA	Volunteers for Economic Growth Alliance
VFTC	Voluntary Fund for Technical Cooperation
WTO	World Trade Organization
YOUNGOs	youth non-governmental organizations

1 'Puzzling, confusing, and … vacuous'

Capacity building from the World Bank to climate governance

Introduction

2016 was the warmest year on record. As was 2015. And 2014. Heated by political fire and extreme weather events, 2017 marks a shift in global approaches to the changing climate. The inauguration of Donald Trump, with his public climate denial and withdrawal from the Paris Agreement, signalled a shift in climate action and diplomacy.

The Paris Agreement, adopted in November 2015 at the 21st Conference of the Parties (COP21) of the United Nations Framework Convention on Climate Change (UNFCCC) marks a vital act in climate change diplomacy. The framework for the implementation of the Paris Agreement was established at the 22nd COP, held in Marrakech, Morocco, in 2016. This was unexpected, when the timeframe for its implementation was put as 2020 – four years later. This shifted the agendas for the following Conferences of the Parties into the 'implementation COPs', to work out what the Paris Agreement really means and how it will actually work. Such a rapid entry into force of the Paris Agreement adds momentum to the global community's efforts to build a zero-carbon, climate-resilient future. Now Parties must adopt procedures for operationalizing the new frameworks, institutions and processes established under the Paris Agreement, and there are many. These include building an 'Enhanced Transparency Framework' to see how countries are doing in reducing emissions (mitigation). It includes sorting out the complex systems for delivering finance to help developing countries deal with climate change. Crucial will be a 'global stocktake' every five years to assess how countries are doing, and building a 12-member compliance mechanism to which Parties can agree. A clearinghouse for risk transfer and insurance will need to be established, a task force will be set up to devise integrated approaches to deal with climate-induced displacement, and a new market mechanism and a global sustainable development mechanism must be created (Dagnet et al. 2016).

All these initiatives will take collaboration and focus through dozens of major and minor meetings of experts and negotiators. But one area of work has been barely discussed: the need to follow through on promises made by the UNFCCC Paris Committee on Capacity Building, adopt a five-year work plan, and develop the Capacity Building Initiative for Transparency (CBIT). The Paris Agreement's decisions to establish the Paris Committee on Capacity Building under Article 11, create the CBIT under Article 13, and in between, to stipulate in Article 12 that countries will promote education, training and public awareness in dealing with climate change, can be regarded as foundational for all the other institutions, mechanisms and processes. To achieve a low-carbon, climate-resilient world, capacity building to reduce greenhouse gas emissions and adapt to its increasing impacts in an open and transparent matter, as stated in the Paris Agreement, is of central importance to facilitate the implementation of all the other provisions and decisions.

This is especially true for the nano-emitters, the least developed countries (LDCs) and Small Island Developing States (SIDS), who are collectively regarded as the particularly vulnerable countries (PVCs), which have been hit first and hardest by the increasing impacts of climate change and have the least capacity to adapt. Subsequent sections of the chapter will explain the salience and argument of the book – why capacity building is important for addressing climate change and how it can be done on a long-term, sustainable basis. We begin by tracing how the concept of capacity building has evolved in the domain of international development cooperation, and then turn to how it is being thought of in humanity's response to climate change. We end this brief introductory chapter with a roadmap for the rest of the book.

A child of the World Bank: where the category of 'capacity building' came from

Looking at the evolution of international development cooperation since the 1950s, capacity building can be said to have its precursor in the concepts of 'institution building', 'institutional strengthening', 'human resource development', 'institutional economics', etc. (Morgan 2006; Kuhl 2009; Keijzer & Janus 2014). Based on the experience of the US-led Marshall Plan to rebuild war-ravaged Europe after the Second World War, the USA and the European countries believed that development could be best pursued in newly decolonized developing countries by building and strengthening their national institutions. But social engineering proved a much more complex phenomenon than physics and mechanics, particularly before and in the initial stage of capitalist development, so thinking began to change. The concept of 'institutional economics' soon evolved as a response to dissatisfaction with traditional technical cooperation, where development of local human capital was not the focus (Thorbecke 2000). The new discipline of 'institutional economics' tried to establish the idea of institution building with some theoretical underpinnings (Booth 2011; Hilderbrand 2002). The argument was that differences in economic growth and development among developing countries can be explained by the differing quality of the institutions responsible for economic management. But the relevance of politics to institutional change remained unappreciated by the aid agencies, thus, as Shirley argued, the new institutional economics was not good news for development assistance (2008: 76).

The concepts of 'capacity development' and 'capacity building' have a relatively weak intellectual pedigree within development theory. Capacity development and capacity building may be regarded as an amalgamation of various precursor concepts, including 'institution building', 'institutional strengthening', 'human resource development', 'organizational development', 'community development', 'sustainable development' and 'institutional economics', among others (Lusthaus et al. 1999; Morgan 2006; Kuhl 2009; Keijzer & Janus 2014). Due to capacity building's evolution as a blend of precursors, different disciplines employ a 'wide range of implicit mental frameworks about capacity', various relevant actors hold diverse perspectives on capacity building work, and there exists 'no broadly accepted definition' of the concept (Morgan 2006). The lack of widely accepted definitional boundaries on 'capacity' has led capacity development to be regarded as an 'umbrella concept' (Morgan 1998) that can be used to link 'previously isolated approaches to a coherent strategy with a long-term perspective and vision of social change' (Lusthaus et al. 1999).

The nature of 'capacity development' as a catch-all term has proved to be both advantageous as an 'integrating force that brings together a large number of stakeholders', as well as disciplines and aid objectives, and disadvantageous due to the danger of the concept being 'used as a slogan rather than as a term for rigorous development work' (ibid.). As approaches to nation-building ranging 'from the macro and the abstract ... to the micro and the operational' may currently be packaged and legitimized as capacity development by donors and civil society organizations and thereby garner broad-based support, there has been little movement among these actors to make the concept of capacity any less 'puzzling, confusing, and ... vacuous' (Morgan 2006).

Examples of different organizations' own definitions of capacity development may aid in illustrating how capacity can be broadly understood and stretched to back a wide variety of aid initiatives. Definitions employed by the United Nations Development Programme (UNDP), the World Bank, the Organization for Economic Cooperation and Development (OECD), and the Canadian, German, and American development agencies are provided in Table 1.1.

Across the many definitions, there appears consensus that capacity building should be driven by local demands and contribute to sustainable development of institutions and societies. However, aside from these broad similarities, it seems that various perspectives on capacity building are

Table 1.1 Definitions of capacity development by selected development agencies

Agency	*Definition*
UNDP	Capacity development: 'the process through which individuals, organizations, institutions and societies develop abilities to perform functions, solve problems and set and achieve objectives' (UNDP 2006)
World Bank	Capacity development: 'A locally driven process of transformational learning by leaders, coalitions and other agents that leads to actions that support changes in institutional capacity areas – ownership, policy, and organizational – to advance development goals' (World Bank)
OECD	Capacity development: 'The process by which individuals, groups and organisations, institutions and countries develop, enhance and organise their systems, resources and knowledge; all reflected in their abilities, individually and collectively, to perform functions, solve problems and achieve objectives' (OECD 2006)
CIDA	Capacity building: 'A process by which individuals, groups, institutions, organizations, and societies enhance their abilities to identify and meet development challenges in a sustainable manner' (CIDA 1996, in Lusthaus et al. 1999)
GIZ	Capacity development: 'The self-driven process through which people, organizations and companies mobilize and build out their capabilities in order to achieve capacity' (GIZ, translation)
USAID	Capacity building: 'An on-going evidence-driven process to improve the ability of an individual, team, organization, network, sector or community to create measureable and sustainable results' in terms of effectively 'apply[ing] its skills, assets and resources to achieve its goals' (USAID 2012)

simultaneously being employed (Lusthaus et al. 1999). One useful system for grouping these different approaches to capacity building, proposed by Lusthaus, applies the categories of organizational, institutional, systems, and participatory-process definitions. Organizational approaches to capacity development see 'an entity, organization or even set of organizations as the key to development' and therefore focus on the capacities of organizations, such as governments, non-governmental organizations (NGOs), and community organizations. Institutional approaches, inspired by institutional economics, understand institutions as 'the formal and informal "rules of the game"' and concentrate on building 'the capacity to create, change, enforce, and learn from the processes and rules that govern society'. Systems approaches regard capacity building as 'a complex intervention that encompasses multiple levels and actors, power relationships and linkages' and suggest that capacity development should not create new systems, but should 'build on what exists in order to improve it'. Finally, participatory-process approaches emphasize the importance of undertaking capacity development within a 'participatory, empowering partnership for which those involved feel a high degree of ownership' (ibid.).

In 1998, the Canadian International Development Agency (CIDA) consultant Peter Morgan cogently argued that 'capacity building is a risky, murky, messy business, with unpredictable and unquantifiable outcomes, uncertain methodologies, contested objectives, many unintended consequences, little credit to its champions and long time lags'. In 2006, he began a study on capacity with the anonymous quote 'I can't define capacity, but I know it when I see it.' Both statements continue to characterize modern approaches to capacity: it remains a hazily defined goal that all actors can agree is important, but more specific formulations of the boundaries, methodologies, and objectives of capacity building remain deficient or even contradictory.

Capacity first emerged as a prominent concept within international relations with the Marshall Plan, under which the United States endeavoured to rebuild war-ravaged Europe after the Second World War. During this period, and increasingly as the Cold War escalated, the USA and various European nations understood efforts to build and strengthen developing countries' national institutions, especially within newly decolonized nations, as important to shore up political stability. The approach to capacity development as an integral aspect of international security was exemplified by the Truman Doctrine, announced in 1947, which pledged American aid to any democratic nation threatened with 'internal or external authoritarian forces' (U.S. State Department, Office of the Historian n.d.).

In the 1950s, development cooperation by industrialized countries was characterized by an 'industrialization-first', technical assistance-focused strategy with little soft technology transfer. In this period, industrialization was regarded 'as the engine of growth which would pull the rest of the economy along behind it', and development aid therefore consisted largely of investment in 'industrial activities and social overhead projects' (Thorbecke 2000). Accordingly, 'economic growth became the main policy objective' of development cooperation and gross national product (GNP) growth was adopted as 'both the objective and the yardstick of development' (ibid.).

From the 1960s till the 1980s, the aim of strengthening institutions and building human capital replaced GNP growth as the primary objective of development aid. This shift in focus towards institutional capacities and human capital evolved in response to the failures of GNP-oriented development strategies to cope with emergent problems, including unemployment, unequal income distributions, high poverty rates, growing urban congestion, and foreign indebtedness, as well as the increasing prominence of endogenous growth theories (ibid.). The emergent governing discipline, based on the endogenous growth school, identified 'low human capital endowment as the primary obstacle to the achievement of the potential scale economies that might come about through industrialization' and therefore prioritized allocation of resources to research and development activities and spillovers of know-how from one firm or industry to others (ibid.).

Another new framework for the development process that rose to prominence in the 1980s and contributed to the modern understanding of capacity building was new institutional economics, based on the concept that 'appropriate institutions and rules of the game are essential to provide pro-development and anti-corruption incentives' (ibid.). Under this framework, differences in economic growth and development among developing countries are attributed to discrepancies in the quality of institutions responsible for economic management. New institutional economics, however, was prone to concentrate on the technical and mechanical aspects of institutional change at the expense of nuanced considerations of political conditions and other local circumstances (Shirley 2008).

Out of the endogenous growth and new institutional economics paradigms dominant in the 1980s emerged the concept of capacity development/capacity building, which became 'the buzzword of development in the 1990s' (OECD 2006). The World Bank is regarded as the initiator of this concept, although other development agencies later started employing the term 'capacity development'. Some commentators find no basic difference between these two terms (Vincent-Lancrin 2007), while others argue that there is: capacity building starts from scratch, while capacity development is a process beginning from an existing base (Kuhl 2009; Pearson 2011a).

Drawing on the endogenous growth school, capacity building emphasized 'the need to build development on indigenous resources, ownership and leadership and by bringing human resources development to the fore' (OECD 2006). Simultaneously, the collapse of the Soviet influence in Europe and Asia and the 'sluggish response to the first generation of economic reforms in Africa' underscored the importance of institutional factors to development, leading to the prioritization of building institutional capacities (Booth 2011). As the development community shifted its focus from an overarching aim of '*assistance* to a less dependent "help yourself" attitude' throughout the 1990s, agencies zeroed in on capacity building/development as the organizing theme for aid efforts (OECD 2006).

Since the turn of the millennium, there have been four high-level deliberations on aid effectiveness: in Rome in 2003; in Paris in 2005; in Accra in 2008; and in Busan in 2011. Beginning in particular with the Paris Declaration on Aid Effectiveness in 2005, 'systems' development became a focus of aid delivery. Over this period, the shibboleth of development cooperation began to shift from 'aid effectiveness' to 'development effectiveness' (Mawdsley et al. 2014). The landscape of development cooperation continues to change, with new donors emerging from both the Global North and the Global South, and more prominent participation of other stakeholders, including civil society groups.

Whatever the case, there is as yet no consensus on what capacity building/development actually means or entails. Most of the aid agencies have defined it in their own way. But there appears to be a consensus that capacity building must include individuals, institutions/organizations and systems that collectively enable effective development. However, based on an increasing number of sociological studies, Kuhl (2009) argues that development assistance can no longer be primarily explained by the needs of developing countries, rather, one has to consider the search for acquiring legitimacy for continuing development assistance within the domestic constituencies of the industrial world. Regardless, capacity building and capacity development appear to have come to stay as the primary terms in the jargon of international development. And now the concept has spread to environmental and climate change governance. Since the UNFCCC uses the term 'capacity building', this book will go with it. We will now turn briefly to chart the breadth and spread of the term in the domain of global environmental governance.

Capacity building is an integral part of global environmental governance

Since 1991/1992, the concept of capacity building has become part and parcel of the five 'Rio Conventions' and almost all other environmental agreements and protocols. The reason is obvious: environmental management, particularly of cross-border pollution problems, was a new phenomenon in the last three decades, when countries, particularly developing ones, did not have any experience in dealing with them. Earlier decisions relating to capacity building were taken by the Commission on Sustainable Development at its fourth (1996), fifth (1997) and sixth (1998) sessions and by the United Nations General Assembly at its Special Session to review the implementation of Agenda 21 (1997). The UN General Assembly Resolution (UN, A/RES/50/120 Art. 22) of 1997 refers to the 'objective of capacity-building' as 'an essential part of the operational activities of the U.N.'. So capacity building for environmental management in developing countries has become part of development cooperation, as displayed below by examples of severalmultilateral environmental agreements with provisions explicitly dedicated to capacity building:

- *The 1987 Montreal Protocol* on protecting the ozone layer in Article 9 called for all its Parties to

> co-operate, consistent with their national laws, regulations and practices and taking into account in particular the needs of developing countries, in promoting, directly or through competent international bodies, research, development and exchange of information on:

- a best technologies for improving the containment, recovery, recycling, or destruction of controlled substances or otherwise reducing their emissions;
- b possible alternatives to controlled substances, to products containing such substances, and to products manufactured with them; and
- c costs and benefits of relevant control strategies (UNEP 2000).

- *The 1992 Earth Summit* recognized capacity building as one of the means of implementation for Agenda 21, the global blueprint for sustainable development. Chapter 37 of Agenda 21 gave particular focus to national mechanisms and international cooperation for capacity building in developing countries (Chapter 37: 1).
- *The UN Framework Convention on Climate Change* (UNFCCC) in Article 6 laid out a plan for promoting education, training and public awareness of climate change. We will return to this later.
- *The Convention on Biological Diversity* (CBD), Articles 13 and 18 discussed the need to build capacity in developing countries, as did the *UN Convention to Combat Desertification* (UNCCD) in its Article 19.
- *The 2001 Stockholm Convention* Article 10 called for 'Public information, awareness and education', by 'Provision to the public of all available information on persistent organic pollutants, and the development of educational and public awareness programmes [and for] Public participation [and] Training of workers, scientists, educators and technical and managerial personnel.'
- *The Nagoya Protocol* on sharing the benefits of biodiversity in its Article 22 agreed that

  > [its] Parties shall cooperate in the capacity-building, capacity development and strengthening of human resources and institutional capacities to effectively implement this Protocol in developing country Parties, in particular the least developed countries and small island developing States among them, and Parties with economies in transition, including through existing global, regional, sub-regional and national institutions and organizations.

  In particular, it called for attention to Parties and groups' 'capacity to implement, and to comply with the obligations of, this Protocol'; their 'capacity to negotiate mutually agreed terms'; their 'capacity to develop, implement and enforce domestic legislative, administrative or policy measures on access and benefit-sharing'; and their capacity 'to develop their endogenous research capabilities to add value to their own genetic resources' (Convention on Biological Diversity n.d.).
- The latest global action agenda, the *Sustainable Development Goals* (SDGs) for 2030, prominently highlights the need for capacity building. SDG target 17.9 of the 2030 Agenda for Sustainable Development

is the dedicated target to capacity building and aims to 'Enhance international support for implementing effective and targeted capacity-building in developing countries to support national plans to implement all the sustainable development goals, including through North-South, South-South and triangular cooperation.' The means of implementation have been agreed to include the mobilization of financial resources as well as capacity building and the transfer of environmentally sound technologies to developing countries on favourable terms, including on concessional and preferential terms, as mutually agreed. Member States also commit respectively in paragraphs 109b and 109c 'to strengthen their national institutions to complement capacity-building' and

> ensure the inclusion of capacity-building and institution-strengthening, as appropriate, in all cooperation frameworks and partnerships and their integration in the priorities and work programmes of all United Nations agencies providing assistance to small island developing States in concert with other development efforts, within their existing mandates and resources.

Among the Means of Implementation listed under Chapter VI of the outcome document of the Rio +20 Conference, called 'The Future We Want', capacity building is the subject of paragraphs 277–280. Member States commit to emphasizing the need for enhanced capacity building for sustainable development and strengthening technical and scientific cooperation, to call for the implementation of the Bali Strategic Plan for Technology Support and Capacity-building, adopted by UNEP and to invite relevant agencies of the UN system and other international organizations to support developing countries, especially least developed countries, in capacity building to develop resource-efficient and inclusive economies.

Looking at the array of these multilateral environmental agreements (MEAs) and the sustainable development goals agenda, it is clear that there exist several synergies that together form a set of generic needs for capacity building across agreements, such as the capacity for the formulation of action plans and programmes to address the problems in question, developing plans for compliance with the MEAs, including periodic reporting obligations, sustainable management of natural resources and participation in MEA negotiations. The UN General Assembly Resolution 69/237 of 19 December 2014 on building capacity for the evaluation of development activities at the country level requires the UN Secretary General to provide an update on progress made in building capacity for evaluation. Across all these initiatives, it has been agreed that LDCs be given preference in efforts to build capacity. However, there are MEA-specific capacity building needs in each case. The next section will focus on the specific provisions and decisions undertaken within the UNFCCC, the focus of this book.

Capacity building under the UNFCCC, the Kyoto Protocol and the Paris Agreement

Though little discussed, capacity building has been a part of negotiations under the UNFCCC since its inception in 1992. Article 6 of the Convention is dedicated to promoting education, public awareness, and public access to climate change information, public participation in addressing climate change, and training of scientific, technical and managerial personnel. In a similar vein, the Kyoto Protocol in its Article 10 (paras d and e) proposes strengthening research capacity, education and training of personnel in developing countries. Below is a brief timeline of capacity building developments under the recent rounds of the UNFCCC:

- The Marrakech Accords at COP7 elaborated the capacity-building agenda, setting a framework, guiding principles, priority areas, with particularly vulnerable country needs to be given preference (UNFCCC 2001).
- In 2011, COP17 created the Durban Forum on Capacity Building (UNFCCC 2011), a multi-stakeholder forum, which meets annually during negotiations to share ideas and best practices relating to climate change; the fifth meeting of the Forum took place in Bonn in May 2016.
- At COP18, in 2012, Parties adopted the eight-year Doha Work Programme on Convention Article 6, which requested an annual in-session Dialogue on Article 6 issues (UNFCCC 2012).
- In 2014, the UNFCCC Secretariat launched a Web portal on capacity-building activities.
- In 2014, at COP20 in Lima, Parties decided to have an annual Ministerial Dialogue on Article 6, to sensitize the political leadership to the issues of education and public awareness (UNFCCC 2014).
- Finally, at COP21, Parties established the Paris Committee on Capacity Building (PCCB) in order 'to address gaps and needs ... including with regard to coherence and coordination in capacity building activities under the Convention' (UNFCCC 2015). The text directs the Subsidiary Body for Implementation to organize annual in-session meetings of the Paris Committee on Capacity Building, develop its Terms of Reference, instructs the Paris Committee on Capacity Building to oversee a work plan for the period 2016–2020; COP21 also establishes the Capacity Building Initiative for Transparency (CBIT) (ibid.), to strengthen national institutions to meet Article 13 provisions.

What is the outcome of all these provisions and decisions in the last decade?

Development cooperation by industrial countries in the form of technical assistance began in the 1950s. Since then, it has appeared under many different names and forms, and this idea of capacity building has gained a certain momentum of its own in development circles and environmental governance negotiations. But what has been accomplished, and what still requires attention?

With long years of experience in attending climate negotiations with the Bangladesh delegation, Khan, a co-author of this book, observed that since 2001 capacity building as a negotiation agenda had almost always been a low-key issue, with no serious disagreements between industrial and developing country delegates. Like motherhood and apple pie, who could be against building the capacity of developing nations to negotiate and manage climate change impacts and plan the shift to the green economy?

Disagreements sometimes surfaced about the role of the UNFCCC in the implementation of capacity building measures. Currently, about 12 organizations within the UNFCCC alone, together with a larger number of other multilateral and bilateral agencies, are involved in capacity-building activities (Dagnet & Northrop 2015). At COP21 in Paris, delegations from the Global North maintained the argument that support for capacity-building initiatives should continue through their national development agencies, and not be taken over by the UN, with substantial control by recipient countries. However, delegates from the Global South have been pushing for a greater UNFCCC role, as there is currently no central agency to ensure coherence and coordination among so many disparate agencies and processes pursuing the capacity-building agenda. The Paris Committee on Capacity Building (PCCB) established under the UNFCCC could potentially play this coordination role, but its budget, authority and remit – in short, its own capacity – are uncertain. Will other existing agencies and parts of the Convention hierarchy and those in development agencies and multilateral bodies give up their own authority and turf to allow the PCCB to effectively coordinate activities? Everyone likes to talk about coordination, but few like to be coordinated, i.e. told what to do.

Second, in the technical assistance programmes of capacity building, private consulting firms – usually from donor countries – have usually been commissioned to do the job. When a lack of capacity is observed in a developing country by an aid agency wishing to help in area X, one or two consultants have typically been 'parachuted in': they organize some workshops and training programmes, the project gets done, and it all ends with the submission of a project report. This renders capacity building mainly an input-based, supply-driven, short-term, and ad-hoc exercise. In this classic model, no capacity building 'systems' were left behind to carry the task forward (Huq 2016). Such donor-driven exercises by foreign experts *might even harm local capacity-building initiatives* and undermine local knowledge and management systems (Godfrey et al. 2002). Capacity building is a long-term iterative affair, but aid agencies have built-in incentives to grind out project completion reports which document short-term output-based results. There is the contrary experience of some countries which have managed an endogenous process of increasing capacities, where development cooperation played a stimulating, but not decisive, role (Kuhl 2009).

There is virtually no research to date on how much money has been spent on work to build capacity in the area of climate change. Some estimates suggest that one-third to one-fourth of annual foreign assistance in different areas might go to some form of capacity building, with an overwhelming share spent by bilateral agencies (Morgan 2006; Victor 2013) Since capacity building as a cross-cutting issue often remains a component of other projects, it is difficult to quantify the total funding specifically dedicated to capacity building for climate change, but even generous estimates would probably conclude that funding for capacity building in this area remains very poor (Chen & He 2013). That is, most work is done as 'projects', without developing the institutional structures in countries that could autonomously handle climate change mitigation and adaptation issues in future years.

Fourth, the private sector is largely absent in capacity-building activities, except perhaps in the insurance sector, which is a direct profit-earning venture (Victor 2013). But the private sector is a major stakeholder in the capacity-building agenda. Significant work is needed to envision and develop ways for private businesses to advance national climate efforts.

Finally, there is a sizeable literature on aid effectiveness, but few such assessments of capacity-building activities. Capacity building involves both software and hardware, as new knowledge, skills and technologies are needed, which can then create an enabling environment for learning and research for individuals and institutions over the long term. Morgan (1998: 6) cogently argued 20 years ago that 'capacity building is a risky, murky, messy business, with unpredictable and unquantifiable outcomes, uncertain methodologies, contested objectives, many unintended consequences, little credit to its champions and long time lags'. It will take a new perspective on the concept, and a new set of institutional incentives, to build real capacity.

Time for a fresh approach to developing a global strategy on capacity building

The question of aid effectiveness has been a concern in development cooperation since it began decades ago. As development engineering in varied soils is often a process of 'learning by doing', themes and strategies for ensuring aid effectiveness have changed consistently. Beginning with institution building and institutional strengthening in the 1960s and the 1970s, aid organizations have zeroed in on capacity building/development since the 1990s, as a strategy for fostering endogenous growth in developing countries through 'good governance' and 'country ownership' of exogenous assistance from development partners. Since 2003, there have been four high-level deliberations on aid effectiveness – in Rome in 2003, in Paris in 2005, in Accra in 2008 and finally in Busan in 2011. Beginning particularly with the Paris Declaration on Aid Effectiveness in 2005, 'systems development' was given focus in aid delivery, but so far very limited evidence is available on how the recipient countries themselves are managing capacity development strategies (Keijzer & Janus 2014). Meanwhile, the shibboleth of development cooperation is shifting from 'aid effectiveness' to 'development effectiveness' since Busan in 2011. As mentioned above, the landscape of development cooperation is also changing, with the arrival of new donors from both the Global North and the Global South, and some stakeholders including increased civil society participation.

In such evolving dynamics, the total number of aid projects/programmes has kept increasing, with hardly any capacity-neutral interventions, but things on the ground have changed little (Keijzer 2013). While some analysts hold both the donors and recipients responsible for this (Wood et al. 2011; Keijzer & Janus 2014), others argue that developing country donors are slow learners (Mawdsley et al. 2014) and lag behind the recipient countries in recognizing the principles of effective aid, such as mutual accountability and transparency (Gulrajani 2014). Obviously, inefficiency and ineffectiveness in capacity-building initiatives continue to linger, as stated in the UNFCCC third review (UNFCCC 2016), mainly because of short-lived project-based interventions by so many actors with little coordination among them.

This book is about overcoming the lingering gaps and lacunae in capacity-building activities exclusively devoted to addressing climate change. We would argue that the main indicators used to judge value for money for climate change capacity building should be whether in-country capacity systems and capacity suppliers have been left behind in each target country. This is how one can see the money being spent as 'investment' rather than as 'expenditure'. Tackling climate change issues requires the building of long-term, sustained systems at national levels to carry out capacity-building functions for decades and generations to come. It is time to highlight the discrepancy between investing for capacity building and simply spending money on consultants.

Here the role of universities comes in. Universities are among the most enduring and sustainable institutions in existence, some being almost a thousand years old, and many have outlived empires and political regimes in some of the most fragile states in the world. Under Articles 11 and 12 of the Paris Agreement, investing in universities to set up and sustain capacity-building systems seems to be a no-brainer option (Hoffmeister et al. 2016).

Every country, even the poorest, has universities which teach environmental science or geography, and they can be brought in to teach climate change issues and train young and not-so-young people for positions in government, non-profits, and the private sector to lead, manage and act directly on climate change. Some universities in developing countries are already taking the lead in developing Master's programmes for students and professionals. Winthrop and McGivney (2016) rightly argue that universities have acted historically as the main arbiters of knowledge and have proved to be powerfully rich soil in societies where the seeds of mass schooling have flourished. So universities have multiplier effects both up and down.

However, most of the universities, especially in the particularly vulnerable countries, lack resources, such as a budget to develop infrastructure, technical aids and even internet facilities for learning; many lack access to global knowledge and databases, have poor library collections, and poor or no research funds. The list goes on. Overcoming these barriers requires funding and appropriate programme development to offer degrees and short-course programmes to impart the specific skills relevant to address climate change. Low-cost, high-impact activities of universities could include global engagement of teachers, research collaborations, access to peer-reviewed knowledge, distance learning, student and faculty exchanges, and so on (Hoffmeister et al. 2016).

Instead of the existing donor-driven workshop and short training-based capacity building programmes led by foreign consultants, we need to create in-country suppliers of capacity building. We would argue that local universities can best play this role. Other knowledge generators, such as private companies, think tanks and NGOs, must join and contribute to this process, but let the universities be the central hub as the coordinator and validator of national capacity-building activities. Wood et al. (2011: xiv) support our point here:

> The complex, long-term challenges of capacity development are the most important constraints for most countries, and these do not allow for 'quick fixes' or bureaucratically engineered solutions. However, partner countries can do more to identify priorities for strengthening capacities in targeted areas. Donors and agencies in turn can do more to support those priorities in coordinated ways, to strengthen country systems by using them and to reduce donor practices that undermine the development of sustainable capacity.

This book is exactly about how to address these problems.

Structure of the book

The book is divided into nine chapters. This introductory chapter lays out the context for the book, providing a snapshot of how capacity building has been used in international development cooperation efforts and in international environmental treaties, regarding climate change in particular. In the process of telling this history we challenge the results/outcomes approach of past and current capacity-building initiatives, upon which donors have spent millions of dollars. What is the end result today? How much capacity and what kind of capacities have been built? Who led the process? Was it demand-driven or supply-driven? Are these efforts adequate to the problem at hand, which is developing the ability of a whole national society to respond to the crisis of climate change? Have any sustainable systems been left in place? If not, what are the gaps and lacunae in standard processes? In this introductory chapter we have argued that with the Paris Committee on Capacity Building established at COP21, it is time to think of a fresh approach to framing a global strategy that can overcome existing inefficiencies and ineffectiveness and devise a long-term sustainable capacity-building system in the developing countries. We suggest local universities as the logical focus of future capacity-building initiatives, and propose taking a new, long-term approach to supporting faculty and staff development, locally-run training programmes and research and educational programme cooperation with universities and agencies around the world.

Chapter 2 traces the evolution of capacity building as the framework of development cooperation since the 1950s, then zeroes in on the agenda of capacity building for climate change since the adoption of the UN Framework Convention on Climate Change (UNFCCC) in 1992 up to COP21 in Paris. In the process, the chapter discusses the evolution of institutions, mechanisms and processes dedicated to promoting the agenda of capacity building, the availability of funding for the purpose, etc. Since decision-making through negotiations among 195 disparate parties under the UNFCCC is very much a political process, this history also teases out the politics and role of actors in the adoption of the main elements and strategies of capacity building in addressing climate change. This chapter highlights the interests of development agencies and consulting firms, NGOs, universities and think-tanks in making capacity building happen in certain ways.

Continuing the historicized approach to capacity building by relevant stakeholders, Chapter 3 first traces experiences with capacity building initiatives in a few other multilateral development and environmental regimes where capacity building has been a significant component of development cooperation. Obviously, learning from their experiences, discovering the similarities and variations among them would be a welcome exercise to chart a new framework of capacity building under the Paris Committee of the UNFCCC. For that purpose, the experiences of capacity building under the regimes of the World Trade Organisation, the Regional Seas Programme in the Baltic and Mediterranean regions, human rights law, disaster risk reduction and the Montreal Protocol have been reviewed.

Taking into account the experiences of these regimes, Chapter 4 is devoted to establishing a capacity-building framework that is up to the task of addressing the challenge of climate change. It first traces the characteristic features of climate change as a complex, long-term, collective action problem, and then proposes capacity building framing elements, strategies and processes at different levels of action – individual, organizational/institutional and systemic. The suggested elements, strategies and processes, including the role of universities in such a framework, look at the demand and supply sides of capacity building and how a demand-driven approach can help create a long-term sustainable system of capacity building. In the process, the issue of ownership of capacity building through the agreed principles of partnership is critically analysed.

Chapter 5 is devoted to understanding the dynamics of climate change impacts in three case study countries: two least developed countries (LDCs), i.e. Bangladesh and Uganda, and a Caribbean small island state, Jamaica. These three country case studies were selected based on the understanding that LDCs' and Small Island Developing States' (SIDS') needs for capacity building are the greatest, and were also undertaken in order to explore realities on the ground in capacity-building activities during the past decade. The chapter analyses country-level information from Bangladesh, Uganda and Jamaica, beginning by presenting the three country profiles, their basic socio-economic parameters, the climate change impacts in these three countries and relevant project profiles, including national adaptation programmes of action (NAPAs) and Intended Nationally Determined Contributions (INDCs) to address the climate problem. A discussion of similarities and differences among capacity-building efforts in the three countries follows.

With findings from the three case study countries analysed, Chapter 6 makes an analytical review of capacity-building activities undertaken by the major bilateral and multilateral agencies, including the World Bank, Global Environment Facility, the Asian Development Bank (ADB),the African Development Bank (AfDB), the UNFCCC Secretariat and its thematic agencies, the Green Climate Fund, and the major bilateral agencies such as AusAid, CIDA, DANIDA, DFID, NORAD, SIDA and USAID. The chapter raises a series of questions about their past spending – how much money has been spent for the purpose, their regional and thematic distributions, whether the money spent so far is regarded as 'investments', or can be written off as 'expenditures', the processes by which spending was carried out and the end result and outcome/impact today. Finally, the chapter lays out the lessons learned from these practices.

The lessons learned from the case study countries and past global practices form the basis on which Chapter 7 establishes the vision of universities as the central hub of capacity building under the Paris Agreement. The rationale for making universities the central hub follows our thesis that universities as higher centres of learning and research are among the most sustainable institutions in the developing countries. This argument is followed by the presentation of findings from the research on some universities in the three case-study countries, including their strengths and weaknesses, their collaboration with external partners, etc. The chapter ends by elaborating a roadmap of what can be done to make universities the core of a new global capacity-building initiative in terms of their infrastructures, funding, programme development and global networking.

Along with Article 13 of the Paris Agreement, which established an enhanced Transparency Framework for Action and Support, COP21 decided to create a Capacity Building Initiative for Transparency (CBIT). Chapter 8 elaborates how the purpose of CBIT, i.e. to build capacity to ensure accountability and transparency of all climate action under the Paris Agreement, especially greenhouse gas (GHG) emissions cuts by developing countries, can be realized. The tools, modalities and guidelines needed to ensure transparency in the monitoring of GHG emissions and climate finance, in both the developed and developing world, are analysed. This is followed by a snapshot of the state of affairs in transparency of actions in two case countries and the opportunities and barriers facing transparency, with a focus on the institutions and funding for capacity building needed to achieve accountability and transparency.

Chapter 9, the concluding chapter, deals with what is required as means of implementation for the capacity building framework proposed in Chapter 4, itself being a crucial means of implementation of the Paris Agreement and also of Sustainable Development Goal (SDG) # 13 on climate change. It elaborates how the proposed framework will address the many elements of the Paris Agreement to which capacity building is crucial, including adaptation and mitigation, the facilitation of technology development and deployment, access to climate finance, the transparent communication of information and relevant aspects of education, training and public awareness. So the chapter focuses on how to implement the framework, with the means and strategies necessary to enhance a long-term sustainable system of capacity building in developing countries to address climate change.

References

Booth, D. (2011). Aid, institutions and governance: What have we learned? *Development Policy Review*, *29*(1), 5–26.

Chen, Z., & Jingjing, H. (2013). Foreign aid for climate change-related capacity building. WIDER Working Paper No. 2013/046, April.

Convention on Biological Diversity. (n.d.). Text of the Nagoya Protocol, Article 22: Capacity. Available at: www.cbd.int/abs/text/articles/default.shtml?sec=abs-22

Dagnet, Y., & Northrop, E. (2015). *3 reasons why capacity building is critical for implementing the Paris Agreement*. Washington, DC: World Resources Institute.

Dagnet, Y., Waskow, D., Elliott, C., Northrop, E., Thwaites, J., Mogelgaard, K., Krnjaic, M., Levin, K., & McGray, H. (2016). *Staying on track from Paris: Advancing the key elements of the Paris Agreement.* Washington, DC: World Resources Institute. Available at: www.wri. org/sites/default/files/Staying_on_Track_from_Paris_-_Advancing_the_Key_Elements_of_the_Paris_Agreement_0.pdf

Godfrey, M., Sophal, C., Kato, T., Piseth, L.V., Dorina, P., Saravy, T., Savora, T., & Sovannarith, S. (2002). Technical assistance and capacity development in an aid-dependent economy: The experience of Cambodia. *World Development, 30*(3), 355–373.

Gulrajani, N. (2014). Organising for donor effectiveness: An analytical framework for improving aid effectiveness. *Development Policy Review*, *32*(1), 89–112.

Hilderbrand, M. (2002). Capacity building for poverty reduction: Reflections on evaluations of UN system efforts. Unpublished MS thesis. Harvard University.

Hoffmeister, V., Averill, M., & Huq, S. (2016). *The role of universities in capacity building under the Paris Agreement Policy Brief.* Dhaka: International Centre for Climate Change and Development (ICCCAD).

Huq, S. (2016). Why universities, not consultants, should benefit from climate funds. Available at: www.climatechangenews.com/2016/05/17/why-universities-not-consultants-should-benefit-from-climate-funds/ (accessed 28 May 2016).

Keijzer, N. (2013). *Unfinished agenda or overtaken by events?: Applying aid and development effectiveness principles to capacity development support.* Bonn: German Development Institute.

Keijzer, N., & Janus, H. (2014). Linking results-based aid and capacity development support conceptual and practical challenges. Discussion Paper 25/2014. Bonn: German Development Institute.

Kuhl, S. (2009). Capacity development as the model for development aid organizations. *Development and Change, 40*(3), 551–557.

Lusthaus, C., Adrien, M., & Perstinger, M. (1999). Capacity development: definitions, issues and implications for planning, monitoring and evaluation. *Universalia* Occasional Paper, *25*(1999), 1–21.

Mawdsley, E., Savage, L., & Kim, S-M. (2014). A 'post-aid world'? Paradigm shift in foreign aid and development cooperation at the 2011 Busan High Level Forum. *The Geographical Journal*, *180*(1), 27–38.

Morgan, G. (1998). *Creative organization theory*. Newbury Park, CA: Sage.

Morgan, P. (2006). *The concept of capacity.* European Centre for Development Policy Management. Available at: http://preval.org/files/2209.pdf (accessed 20 September 2016).

OECD. (2006) Applying strategic environmental assessment: Good practice guidance for development co-operation. Available at: www.oecd.org/environment-development.37353858.pdf

Pearson, J. (2011a). The core concept, Part I. Available at: www.Lencd.org/learning

Pearson, J. (2011b). *Creative capacity development: Learning to adapt in development practice.* Westhart, CT: Kumarian Press.

Shirley, M. (2008). *Institutions and development: Advances in new institutional analysis.* Cheltenham: Edward Elgar.

Thorbecke, E. (2000) The development doctrine and foreign aid 1950–2000. In F. Tarp (ed.), *Foreign aid and development. Lessons learnt and directions for the future* (pp. 17–47). London: Routledge.

UNEP. (United Nations Environment Programme) (2000). The Montreal Protocol on Substances that Deplete the Ozone Layer. Available at: http://unep.ch/ozone/pdf/Montreal-Protocol2000.pdf

UNFCCC. (2001). Framework for capacity building in developing countries. Decision 2/CP7, paras 1–13 plus Annex.

UNFCCC. (2011). Outcome of the work of the Ad-hoc Working Group on Long Term Cooperative Action under the Convention. Decision2/CO.17, Section VI.

UNFCCC. (2012). Doha Work Programme on Article 6 of the Convention. Decision15/CP.18.

UNFCCC. (2014). Lima call for climate action. CP.20. Available at: https://unfccc.int/files/meetings/lima_dec_2014/application/pdf/auv_cop20_lima_call_for_climate_action.pdf

UNFCCC. (2015a). Adoption of the Paris Agreement. Decision 1/CP.21, para 72.

UNFCCC. (2015b). Adoption of the Paris Agreement. Decision 1/CP.21, para 85.

UNFCCC. (2016). Decision-/CP.22. Third Comprehensive Review of the Implementation of the Framework for Capacity-Building in Developing Countries under the Kyoto Protocol. Available at: http://unfccc.int/files/meetings/marrakech_nov_2016/application/pdf/cmp12_i10_3rd_comprehensive_review_for_kp_cb.pdf

UNFCCC Paris Committee on Capacity-building (2017). First meeting of the Paris Committee on Capacity-building. Bonn, Germany, 11–13 May 2017. PCCB/2017/1/10 16 June 2017.

U.S. State Department, Office of the Historian (n.d.). The Truman Doctrine, 1947. Available at: https://history.state.gov/milestones/1945-1952/truman-doctrine (accessed 12 December 2016).

Victor, D. (2013). Foreign aid for capacity-building to address climate change: Insights and application. WIDER Working Paper No. 2013/084.

Vincent-Lancrin, S. (2007). Developing capacity through cross-border tertiary education. In OECD-World Bank, *Cross-border tertiary education: A way towards capacity development* (pp. 47–102). Paris: OECD.

Winthrop, R., & McGivney, M. (2016). Why wait 100 years? Building the gap in global education. Washington, DC: Brookings Institution. Available at: www.brookings.edu/research/why-wait-100-years-bridging-the-gap-in-global-education/

Wood, B., Betts, J., Etta, F., Gayfer, J., Kabell, D., Ngwira, N., Sagasti, F., & Samaranayake, M. (2011). *The evaluation of the Paris Declaration, Phase 2, final report*. Copenhagen: Danish Institute for International Studies.

2 The meagre history and politics of capacity building under the UNFCCC

Introduction

This chapter walks through all the language adopted at the original 1992 Rio Earth Summit on capacity building, and at the 23 Conferences of the Parties (COP) and meetings of the Subsidiary Bodies since then. Together it composes the first comprehensive history of this type of capacity building under the UN Framework Convention on Climate Change (UNFCCC). Beginning with Article 6 of the original Framework Convention, which identified the need for capacity building, developed countries agreed to financially support capacity building in developing countries. Capacity building efforts repeatedly emphasized in the Convention's founding text included research, systematic observation and data-sharing capacities; international research networks; climate-related education, training, and public awareness.

The decades of meetings since have included many calls for more coordination and support for capacity building, and specified the kinds of work that is needed. However, the extent to which these calls were ever heeded remains unclear. Many of these same elements continued to be emphasized up to the 2015 Paris Agreement, when the issue finally attracted a bit more attention. The Paris outcomes were particularly significant in that they included a whole article on capacity building. Notably, the Agreement included the creation of a coordinating Paris Committee on Capacity Building (PCCB) and the Capacity Building Initiative on Transparency (CBIT). However, as the normally uncontroversial issue of capacity building became more contentious in Paris, politics – especially in terms of clashes between the Global North and the Global South – precluded a number of further developments. We discuss these at the end of this chapter, and in those to come.

Capacity building under the UNFCCC, 1992–1994: creation of the UNFCCC

From the time of the drafting of the text of the Convention in 1992 and the Convention's entry into force in 1994, the Parties to the Convention have laid the foundation for the myriad capacity-building efforts that have since proceeded under the UNFCCC.

Article 4 of the Convention's founding documentestablishes an overarching basis for capacity building, connecting the imperative that developing countries implement the terms of the Convention with developed countries' provision of support, in the form of funding, technology and know-how. First, the Article lays out various actions that the Parties must undertake in order to implement the Convention, stating they shall

- develop and publish national emissions inventories and mitigation programmes;
- cooperate in the application of mitigation technologies;
- engage in sustainable management of emissions reservoirs and sinks;
- develop adaptation plans;
- promote and cooperate in climate-related research;
- encourage the widest possible participation in climate-related education, training, and public awareness efforts (UNFCCC, Art. 4, para. 1).

Next, Article 4 stipulates that developed country Parties shall provide developing Parties with 'new and additional financial resources to meet the agreed full costs incurred by developing country Parties in complying with their obligations'. It states also that developed countries shall assist developing countries that are particularly vulnerable to climate impacts in meeting adaptation costs (Art. 4, para. 4) and shall promote, facilitate, and finance the transfer of relevant technologies and know-how to developing country Parties (Art. 4, para. 5). In this process of technology and knowledge transfer, according to the Convention text, developed Parties shall 'Support the development and enhancement of endogenous capacities and technologies of developing country Parties'.

In the final capacity-relevant provision of Article 4, the text asserts that 'the extent to which developing country Parties will effectively implement their commitments under the Convention will depend on the effective implementation by developed country Parties of their commitments under the Convention related to financial resources and transfer of technology' (Art. 4, para. 7).

Article 5 is devoted to the topic of climate-related research and observation. It declares that in carrying out Article 4 commitments, Parties shall develop and support intergovernmental research networks and other international efforts to conduct, assess and fund data collection and other research. Furthermore, the Parties shall support international efforts to strengthen systematic observation of climate data and enhance 'national scientific and technical research capacities and capabilities, particularly in developing countries' (Art. 5). Accordingly, the Parties also agreed to expand access to and promote exchange of the data and analyses obtained. Finally, the article states that developed countries will cooperate to improve developing countries' endogenous capacities to participate in the international research efforts discussed (Art. 5).

Although Article 6 does not use the term 'capacity building', it is dedicated to the capacity building strategies of climate-related education, training, and public awareness, including public access to information on climate impacts and public participation in developing responses to climate change. The article states that Parties will 'promote ... the development and implementation of education and public awareness programmes, ... public access to information on climate change and its effects, public participation in addressing climate change, ... and training of scientific, technical and managerial personnel', through both 'the strengthening of national institutions and the exchange or secondment of personnel to train experts in the field, in particular for developing countries'.

Article 11 of the Convention provides for use of a financial mechanism to disburse funding, including for technology transfer, on a grant or concessional basis, under the guidance of the Conference of the Parties to the Convention (COP). The article does not enumerate the specifics of the mechanism, including its policies, programme priorities, eligibility criteria, or modalities for decision-making, leaving these items to be addressed by the COP at its first session (Articles 11.1, 11.4). Article 21 states that the Global Environment Facility 'shall be the international entity entrusted with the operation of the financial mechanism referred to in Article 11 on an interim basis', which Decision 3 at COP4 amended to a permanent basis (Art. 21, para. 3). We summarize this period in Table 2.1.

Financing capacity building, 1995–2000: COP1 to COP6

From 1995 to 2000, movement on the issue of capacity building first took the form of decisions on the financial mechanism of the UNFCCC, the Global Environment Fund, that would fund capacity building (among other climate-related efforts); second, decisions regarding funding for and implementation of technology transfer and research; and least frequently, decisions devoted to capacity building more generally. Texts from COP1 also established the institutional context of capacity building under the

Table 2.1 Summary of developments at the founding of the Convention, 1992–1994

- Capacity building given strong conceptual foundation in the Convention as necessary for developing countries to fulfil their obligations
- Developed countries agree to financially support capacity building in developing countries
- Capacity building-related topics with significance repeatedly emphasized in the Convention's text include: research, systematic observation, and data sharing capacities; international research networks; climate-related education, training, and public awareness
- Financial mechanism created; GEF made an interim operational entity

Convention, stating in Annex I to Decision 6 that the Subsidiary Body for Scientific and Technological Advice (SBSTA) would 'provide advice ... on ways and means of supporting endogenous capacity-building in developing countries', including advice on education programmes, human resources, training, and how to promote international initiatives relating to research, education, public awareness and capacity building.

The large numbers of decisions relating to finance for capacity building (11/CP.1, 12/CP.1, 11/CP.2, 12/CP.2, 13/CP.2, 11/CP.3, 12/CP.3, 2/CP.4, 3/CP.4) and technology transfer (13/CP.1, 7/CP.2, 9/CP.3, 4/CP.4, 14/CP.4, 5/CP.5, 9/CP.5) and those otherwise relevant to capacity building (5/CP.4, 7/CP.4, 15/CP.4, 10/CP/5, 12/CP.5) from this period prove that capacity building's uncontroversial nature had already begun to establish itself in the Convention's early years, through fledgling pursuits of various Convention objectives. They also suggest that funding was short.

The overarching connection established in the Convention text between support provided for capacity building by developed countries and implementation of the Convention in the developing world also meant that capacity building was highly relevant to many decisions establishing bodies and modalities to achieve the Convention's objectives. In fact, in decision texts from COP1 through COP6, 'capacity', in the context of capacity building efforts on climate-related issues, was mentioned 94 times (for comparison: 'emissions' were mentioned 202 times and 'adaptation' 90 times).

In addition, due to the well-established conceptual significance of capacity building to the Convention's objectives, it became clear in this period that capacity building would necessarily be woven into the various programmes created to implement the Convention. For example, the Clean Development Mechanism (CDM), established as part of the 1997 Kyoto Protocol at COP3, allowed Annex I Parties to undertake mitigation projects in non-Annex I countries and count the resultant emissions reductions against their own emissions to comply with the Kyoto mitigation commitments. Capacity building was relevant to the establishment of the CDM in two ways. First, extensive capacity building activities were important to the successful implementation of the CDM: developing countries' capacities to hold up their ends of institutional linkages with developed country governments, identify useful projects, assess short- and long-term costs and risks, effectively negotiate on projects, set baselines for project evaluation, monitor project activities, and acquire and share data on required enhancement (Annex to Decision 10/CP.5). Second, CDM projects themselves were to consist largely of the transfer of mitigation technologies, and had the potential to have a lasting effect on developing countries' capacities to mitigate emissions by bringing not only hard technologies, but also ideas and know-how for further mitigation. Table 2.2 presents a summary of developments during 1995–2000.

Table 2.2 Summary of developments during 1995–2000

- SBSTA instructed to provide advice on capacity building
- GEF entrusted with operation of the financial mechanism
- Capacity building and technology transfer widely acknowledged in decisions, but definition and scope remain nebulous
- Kyoto Protocol's flexibility mechanisms both require and facilitate capacity building

A framework: the 2001 Marrakech Accords

At COP7, in Marrakech, Morocco, the Parties agreed on a framework that has since guided capacity building efforts in developing countries under the Convention and an analogous framework for efforts in countries undergoing the transition from Communist to free market economies (Dagnet et al. 2015).

These two frameworks sought to clarify the Convention's operative understanding of the scope of capacity building and of capacity-building priorities among Party groups (UNFCCC n.d.). They also aimed to provide guidance for various stakeholders, including donor and recipient Parties, bilateral and multilateral agencies, other relevant government agencies and institutions, and entities under the Convention, such as thematic bodies and the operating entities of the financial mechanism (Dagnet et al. 2015).

The frameworks are based on an understanding that 'there is no "one size fits all" formula for capacity building', as capacity building must instead be 'country-driven' and reflect countries' 'specific needs and priorities' (CP.7/Decision 2, Annex, Art. 5). Capacity building is defined in the Marrakech Accords as 'a continuous, progressive, and iterative process' that should involve 'learning by doing' and allow national institutions, coordinating mechanisms and focal points in developing countries to themselves ensure coordination of capacity building efforts by incorporating traditional skills, knowledge and practices into projects (Articles 6, 10, 11, 12).

The framework for capacity building in developing nations focuses primarily on the need to construct enabling environments within each country and establish national institutions with understandings of climate-related issues and possible responses. These aims necessitate enhancing abilities to collect and interpret data on climate and vulnerability and, accordingly, pursue appropriate mitigation options, create adaptation plans, and act as full members of the Convention by using this understanding to produce submissions, coordinate with other Parties and participate effectively in negotiations. Reflecting its own assertions that capacity-building efforts must be nuanced according to countries' circumstances, the framework includes special priority areas for Least Developed

Table 2.3 The Marrakech capacity-building frameworks (2001)

Developing countries framework: scope of capacity-building needs and areas	*Specific priority areas for LDCs and SIDS*
1 Institutional capacity building: national climate change secretariats or focal points 2 Enhancement and/or creation of enabling environments 3 National communications 4 National climate change programs 5 GHG inventories, emission database management, emissions data systems 6 Vulnerability and adaptation assessment 7 Implementation of adaptation measures 8 Assessment for implementation of mitigation options 9 Research and systematic observation, including meteorological, hydrological, climatological services 10 Development and transfer of technology 11 Improved decision-making, including assistance for international negotiations 12 Clean development mechanism 13 Needs arising out of the implementation of Article 4, paragraphs 8 and 9, of the Convention* 14 Education, training and public awareness 15 Information and networking, i.e. establishment of databases	1 Establishing/strengthening national climate change secretariats or focal points to enable effective implementation of the Convention and participation in the Kyoto Protocol, including preparation of national communications 2 Developing an implementation programme that accounts for the role of research and training in capacity building 3 Developing/enhancing technical capacities to assess vulnerability and adaptation, integrate assessments into sustainable development programmes, and develop national adaptation programmes of action (NAPAs) 4 Establishing/strengthening national research and training institutions to ensure the sustainability of capacity-building programmes 5 Strengthening the capacity of meteorological and hydrological services to collect, analyse, interpret and disseminate weather and climate information to support implementation of NAPAs 6 Enhancing public awareness

Note

* Article 4, paragraphs 8 and 9 of the Convention state that Parties shall consider necessary funding, technology transfer, and insurance actions to meet developing countries' climate-related needs (paragraph 8) and shall account for the special situations of the LDCs in these considerations (paragraph 9).

Countries and Small Island Developing States. Table 2.3 summarizes the Marrakech capacity-building frameworks.

The COP7 frameworks have guided capacity-building activities under the Convention since their establishment. The first comprehensive review of these frameworks was undertaken at COP10 in 2004, and the second and third comprehensive reviews were conducted in 2011 and 2016.

Outside of the frameworks, the Marrakech Accords included a number of other provisions related to capacity building. First, the Parties decided that the Subsidiary Body for Implementation would 'regularly monitor the progress of the implementation' of the frameworks (Art. 10). Second, the Parties recommended that in their first meeting, the Parties to the Kyoto Protocol 'adopt a decision containing a framework on capacity building that reaffirms the [Marrakech] framework ... with additional reference to priority areas for capacity building relating to the implementation of the Kyoto Protocol' (Art. 13). Table 2.4 summarizes the developments at the Marrakech COP7 (2001).

A five-year plan: the 2002 New Delhi Work Programme

COP8 adopted the five-year New Delhi Work Programme on Article 6 of the Convention and slated it for review in 2007 (Decision 11/CP.8). The New Delhi Work Programme recognized that capacities to 'implement Article 6 activities will vary among countries', that 'lack of financial and technical resources could inhibit some Parties' efforts to implement [Article 6] activities', and that it was 'important to learn more from countries regarding the needs and gaps in their Article 6 activities' to enable better targeting of support (Articles 2–5). The Parties agreed that the New Delhi Work Programme would consist of a country-driven, cost-effective, interdisciplinary, holistic, systematic and synergistic approach to Article 6 activities (Art. 9).

The scope of the work programme was defined in four categories of activities that the Parties were encouraged to undertake: (1) international cooperation on activities relevant to the work programme, including cooperation with intergovernmental and non-governmental organizations; (2) promotion and development of climate-focused education and training programmes, targeting youth and including exchange of personnel to train experts; (3) promotion and development of climate-focused training programmes for scientific, technical and managerial personnel; and (4) development of public awareness programmes on climate change, facilitation of public access to information on climate change and promotion of public participation in addressing climate change (Articles 11–14).

Table 2.4 Summary of developments at the Marrakech COP7 (2001)

- Frameworks for capacity building, which had guided relevant activities under the UNFCCC for the past 15 years, were created for developing countries (and nuanced for LDCs and SIDS) and for economies in transition
- The Subsidiary Body for Implementation (SBI) was given the responsibility of reviewing framework implementation, but no Convention body was given a continuous coordination or oversight role

The work programme states that, to implement Article 6 and the New Delhi Work Programme, the Parties 'could, inter alia':

- Develop capacity to identify gaps and needs for the implementation of Article 6, assess the effectiveness of Article 6 activities and consider their linkages to other policies.
- Prepare assessments of needs specific to national circumstances in Article 6 implementation, including use of surveys to determine target audiences and potential partnerships.
- Designate, support and assign specific responsibilities to a national focal point for Article 6 activities.
- Designate and support a national focal point for Article 6 activities.
- Develop a directory of organizations and individuals with indication of their experience and expertise relevant to Article 6 activities, with a view to building active networks for implementation.
- Develop criteria for identifying and disseminating information on good practices for Article 6 activities.
- Increase availability of copyright-free, translated climate change materials.
- Enhance efforts to develop and use curricula and teacher training focused on climate change as methods to integrate climate change issues at all educational levels and across disciplines.
- Seek opportunities to disseminate widely relevant information on climate change.
- Seek input and public participation, including by youth, in formulating and implementing efforts to address climate change; encourage participation by representatives of all stakeholders and major groups in climate negotiations.
- Inform the public about the causes of climate change and sources of GHG emissions.
- Share the findings contained in national communications, action plans, and other programmes with the general public and all stakeholders (Art. 15).

The New Delhi Work Programme also states that, when endeavouring to implement Article 6 activities, the Parties should 'seek to enhance cooperation and coordination at international and regional levels, including the identification of partners and networks with other Parties, intergovernmental and non-governmental organizations, the private sector, state and local governments, and community-based organizations' in order to facilitate the exchange of information, experience and good practices (Art. 16).

The work programme also invites intergovernmental and non-governmental organizations to continue supporting efforts to implement Article 6 and to strengthen and enhance, respectively, existing collaborations relating to these efforts (Articles 17, 18). Parties are also encouraged to forge partnerships with other Parties, intergovernmental organizations, non-governmental organizations, and other relevant stakeholders 'to facilitate the implementation of these activities, including the identification of priority areas for support and funding' (Art. 19). The Parties also recognized that 'the implementation of the work programme will require the strengthening of national institutions and capacities, in particular in developing countries' (Art. 20). Table 2.5 summarizes the developments at the New Delhi COP8 (2002).

The first review of implementation of the Marrakech framework and the first meeting of the Parties to the Kyoto Protocol

In 2004, the first comprehensive review of the implementation of the framework for capacity building in developing countries took place. The COP decided that the framework was still relevant, but that there were significant gaps that needed to be accounted for in continuing implementation efforts (Decision 2/CP.10). Actions designated for increased future attention included:

- prioritizing institutional capacity building;
- raising awareness of climate change at various levels and increasing involvement of national governments in capacity building;
- developing exchange of best practices, experiences, and information on capacity-building activities, including financial resources, case studies and tools for capacity building;
- ensuring that national communications and national adaptation programmes of action provide a good measure of successful capacity building;
- ensuring long-term sustainability of capacity building through its integration into decision-makers' priorities, policies and planning processes;
- making financial and technical resources available, through an operating entity of the financial mechanism, multilateral and bilateral agencies, and the private sector;
- further applying learning-by-doing approaches for capacity building at the national and local levels;
- improving international donor coordination;
- ensuring resources are made available for capacity-building activities.

Table 2.5 Summary of developments at the New Delhi COP8 (2002)

• Established the New Delhi Work Programme on Article 6 of the Convention, recognizing existing inequality among Parties' capacities to implement Article 6
• Characterized by an emphasis on cooperation and coordination among Parties and organizations
• Review of programme planned for 2007

At the first meeting of the Parties to the Kyoto Protocol, it was decided that the Marrakech framework for capacity building would also be used to 'guide capacity-building activities relating to the implementation of the Kyoto Protocol'. Special mention was given to application of the framework to 'enhance the ability of developing countries to participate effectively in project activities under the Clean Development Mechanism', such as institutional capacity building to assist in the designation of national authorities, increasing awareness and training of national authorities and stakeholders, particularly to develop skills related to the CDM project cycle, supporting cooperation and networking between relevant CDM authorities, and enhancing capacity to formulate and implement mitigation policies (Decision 29/CMP.1, Articles 1, 2). Table 2.6 summarizes the developments at COP10 (2004) and COP11 (2005).

Regionalizing capacity building: amending the New Delhi Work Programme

In 2007, the Parties amended the New Delhi Work Programme on Article 6 and extended it for five years, planning another review for 2012 (Decision 9/CP.13, Articles 1, 2). The observations prefacing the Work Programme, as well as the Programme's purposes and guiding principles were unchanged from the original New Delhi Work Programme.

The scope of the amended New Delhi Work Programme is laid out in six activity areas, expanded from the original four areas of international cooperation, education, training, and amalgamated public awareness issues. The first three focus areas are unmodified, but each element in the fourth focus area of the original New Delhi Work Programme is now designated as its own category of action useful 'to advance implementation of Article 6' (Articles 10–16):

- *Public awareness* – Developing public awareness programmes on climate change and its effects from sub-regional to international levels by 'encouraging contributions and personal action ..., supporting climate-friendly policies and fostering behavioural changes, including by using popular media' (Art. 13).

Table 2.6 Summary of developments at COP10 (2004) and COP11 (2005)

• First comprehensive review of the implementation of the Marrakech framework for capacity building in developing countries was undertaken
• Framework deemed still relevant; Parties identified key factors to be taken into account in further implementation
• The first meeting of the Parties to the Kyoto Protocol adopted the Marrakech framework to guide capacity-building efforts in developing countries under the Protocol

- *Public access to information* – Facilitating 'public access to data and information, by providing the information on climate change initiatives, policies and results of action that is needed by the public ... to understand, address, and respond to climate change' (Art. 14).
- *Public participation* – Promoting 'public participation in addressing climate change ..., by facilitating feedback, debate and partnership in climate change activities and in governance' (Art. 15).

To its original list of actions the Parties could take as part of efforts to implement Article 6, the amended New Delhi Work Programme adds just two items (Art. 17):

- Prepare a national Article 6 plan of action, which could be structured according to some or all of the six focus areas, with each one linked to a primary goal and suggested activities (perhaps with defined timeframes and milestones), targets (perhaps specific population groups), and actors.
- Conduct surveys to establish a baseline of public awareness to serve as a foundation for further work and support monitoring of activities' impacts.

A new list of actions that could be undertaken by the Parties to strengthen regional and international efforts was added to the amended New Delhi Work Programme (Art. 19). These actions include:

- promoting awareness of regional and sub-regional needs and concerns;
- strengthening existing regional institutions and networks;
- promoting and encouraging regional programmes and projects that support Article 6 implementation and promoting sharing of data and information, including best practices and lessons learned;
- creating regional portals for the climate change information clearing house (Climate Change Information Network), in collaboration with regional centres;
- developing regional programmes and activities, including training and education materials, using local languages;
- conducting regional and sub-regional workshops to promote exchange of experiences and best practices, and transfer of knowledge and skills.

Articles on support for Article 6 efforts and on intergovernmental and non-governmental organizations are also unchanged in the amended New Delhi Work Programme, apart from the deletion of 'establishment of a mechanism to provide and exchange information' as a priority area for provision of support (Decision/CP.8/Annex/Art. 20). The decision text to which the amended New Delhi Work Programme is annexed is not changed in any significant way from the decision text accompanying the original New Delhi Work Programme (Decision 9/CP.13). Table 2.7 summarizes the developments at COP13 (2007).

Table 2.7 Summary of developments at COP13 (2007)

• New Delhi Work Programme on Article 6 of the Convention amended, extended for five years • Additions to the New Delhi Work Programme included separation of three conceptual areas on which to focus efforts, as well as further suggestions of specific actions Parties could undertake, particularly at the regional and international levels

Copenhagen and Cancun: capacity building through the rockiest years

At COP15, capacity building was not addressed at length or in a decision of its own, as it had been at many preceding COPs, but instead was internalized in the Copenhagen Accord as an important means to support adaptation efforts in developing countries. In the Accord, the Parties agreed that 'developed countries shall provide adequate, predictable, and sustainable financial resources, technology and capacity-building to support the implementation of adaptation action in developing countries' (Decision 11/CP.15, Art. 3). This tenet of the Accord paved the way for the inclusion of capacity building in a similar manner in the Cancun Adaptation Framework the following year.

At COP16, the Parties adopted the Cancun Adaptation Framework, which aimed to emphasize adaptation by establishing that it should be prioritized as highly as mitigation (UNFCCC 2011a). Within the Framework, all Parties were invited to enhance action on adaptation by undertaking 'Research, development, demonstration, diffusion, deployment, and transfer of technologies, practices and processes, and capacity-building for adaptation, with a view to promoting access to technologies, in particular in developing country Parties' (Decision 1/CP.16, Art. 14). Developed country Parties were also requested

> [to] provide developing country Parties, taking into account the needs of those that are particularly vulnerable, with long-term, scaled-up, predictable, new and additional finance, technology and capacity-building ... to implement urgent, short-, medium-, and long-term adaptation actions, plans, programmes and projects at the local, national, subregional and regional levels.
>
> (Art. 18)

Finally, the Parties decided to 'enhance reporting in the national communications of Parties included in Annex I to the Convention on ... the provision of financial, technological and capacity-building support to developing country Parties' (Art. 40).

The Cancun Agreements also internalized capacity building in the context of nationally appropriate mitigation actions (NAMAs) undertaken by developing countries, 'Recognizing that developing country Parties ... could enhance their mitigation actions, depending on provision of finance, technology and capacity-building support by developed country Parties' and deciding to set up a registry of NAMAs 'seeking international support' in order to 'facilitate matching of finance, technology, and capacity-building support for these actions' (Decision 1/CP.16, Section B, Art. 53).

Finally, in a section of the Cancun Agreements devoted to capacity building (Articles 130–137), the Parties acknowledged that capacity building is 'an integral part of enhanced action on mitigation, adaptation, technology development and transfer, and access to financial resources' and decided that funds for enhanced capacity building in developing countries should be provided by Annex II parties and 'other Parties in a position to do so', as well as 'various bilateral, regional, and other multilateral channels' (Art. 131). The Parties also decided that capacity-building support for developing countries should be 'enhanced with a view to strengthening endogenous capacities' by:

- strengthening relevant institutions at various levels, including national coordinating bodies and focal points;
- strengthening information and knowledge networks, including North-South, South-South, and triangular cooperation;
- strengthening climate change communication, education, training, and public awareness at all levels;
- strengthening integrated approaches and the participation of stakeholders in relevant policies and actions;
- supporting capacity building needs in the areas of mitigation, adaptation, technology development and transfer, and access to finance (Art. 130).

Finally, the Cancun Agreements encouraged developed and developing country Parties to continue to report, through their national communications and other submissions, on the support for capacity building activities they give and receive, respectively (Articles 132–135). Table 2.8 summarizes developments at COP15 (2009) and COP16 (2010).

Table 2.8 Summary of developments at COP15 (2009) and COP16 (2010)

• Capacity building internalized in decision on new and additional finance supporting adaptation action, paving the way for inclusion in the Cancun Adaptation Framework
• The Cancun Adaptation Framework internalized capacity building as an important means to achieve enhanced adaptation action and undertake nationally appropriate mitigation actions (NAMAs)
• Importance of capacity building to implementation of the Convention in developing countries was reaffirmed

The Durban Forum and the Doha Work Programme on Article 6

At COP17 in Durban, South Africa, the Parties decided to create a forum for 'in-depth discussion on capacity-building' that meets yearly (during SBI negotiating sessions) in order to enhance the Convention's 'monitoring and review of the effectiveness of capacity-building' (Decision 2/CP.17, Art. 144). The forum aims to fill information gaps resulting from the lack of coordination between stakeholders by giving the Parties, the representatives of Convention bodies, and 'relevant experts and practitioners' a venue to share experiences, ideas, best practices and lessons learned (Art. 144). The decision text establishing the Durban Forum also calls for the publication by the secretariat of a number of synthesis reports to inform the dialogue at meetings of the Forum, including a compilation of information provided by Annex I Parties on progress made in capacity building, a summary of information provided by non-Annex I countries on capacity-building activities, and a compilation of information on capacity building from relevant Convention bodies and international and regional organizations (Articles 148–150). The Forum's inaugural meeting took place in May 2012 in Bonn, at the 36th session of the SBI.

The Durban COP also concluded the second comprehensive review of implementation of the framework for capacity building in developing countries and initiated the third, to be completed at COP22 (Decision 13/CP.17). The Parties decided that the Marrakech framework should continue to guide implementation of capacity building in developing countries (Art. 1) and invited:

- Relevant UN agencies and intergovernmental organizations to continue supporting capacity-building efforts, 'emphasizing and stressing the need for the full involvement of developing countries' in their activities (Art. 3).
- Annex II Parties, other Parties in a position to do so, multilateral, bilateral and international agencies, and the private sector to continue financially supporting capacity-building efforts in developing countries (Art. 4).
- Parties to enhance reporting on best practices in relevant documents (Art. 5).

Finally, the Parties decided that further implementation of the Marrakech framework should be improved by the following activities (Art. 6):

- ensuring consultations with stakeholders throughout the entire capacity-building process (design, implementation, monitoring, evaluation);
- enhancing integration of climate change issues and capacity-building needs into national development plans and budgets;

- increased country-driven coordination;
- strengthened networking and information sharing among developing countries, especially through South-South and triangular cooperation.

At COP18, having considered information prepared under the amended New Delhi Work Programme, the Parties adopted a new work programme on Article 6, to be reviewed in 2020 (Decision 15/CP.18, Articles 1, 2). They also requested that the SBI organize an annual in-session dialogue focused on implementation of the work programme, where the Parties, the representatives of Convention bodies, and relevant experts, practitioners, and stakeholders would share their experiences and exchange ideas, best practices and lessons learned (Art. 9). The Parties planned the in-session dialogue to begin at SBI 38 and alternate focus on an annual basis between two Article 6 conceptual areas: (1) education and training; and (2) public participation, awareness and access to information, with international cooperation considered a cross-cutting theme of both focal areas (Art. 10).

The work programme encouraged the Parties to undertake activities under the same six categories given in the amended New Delhi Work Programme (Decision 15/CP.18, Art. 15). The new work programme, however, also added significantly to the amended New Delhi Work Programme's list of actions that Parties could take to implement Article 6. New implementation activities suggested included:

- Developing communication strategies on climate change based on targeted social research to create behavioural changes.
- Strengthening national education and training/skills development institutions.
- Integrating social networks into Article 6 strategies, i.e. developing social media programmes.
- Supporting informal climate change education.
- Developing tools to support climate change training through collaborative efforts and providing training for groups with a key role in communication and education, including journalists, teachers, youth, children and community leaders.
- Fostering participation of all stakeholders in implementation of Article 6 and inviting their reports on implementation; especially enhancing participation of youth, women, civil society organizations and the media.
- Participating in the annual dialogue on Article 6 under the SBI.
- Seeking to enhance cooperation on Article 6 activities internationally and regionally, including by identifying partners and networks with other Parties, intergovernmental organizations (IGOs) and NGOs, the private sector, state and local governments and community-based organizations.

Similar lists of implementation actions that the Parties could undertake to strengthen regional efforts are included in Article 23. Of the eight included proposals, two are new in the Doha Work Programme:

- Promoting implementation of pilot projects on Article 6 elements through collaborative actions at the regional and national levels, and supporting replication, expansion and sharing of lessons learned.
- Strengthening North-South, South-South and triangular collaboration in matters of climate change-related education, training and skills development.

The new work programme also invited UN organizations to continue supporting Article 6 implementation efforts, including by providing financial and technical support, strengthening collaboration with other intergovernmental organizations, mobilizing partnerships and networking among the Parties, IGOs, NGOs, academia, the private sector, state and local governments, and community-based organizations; promoting in concert with the Parties and civil society the organization of Article 6-focused workshops at all levels, and participating in the annual Article 6 dialogue under the SBI (Art. 24).

Non-governmental organizations were, in turn, encouraged to continue their Article 6-related activities while considering ways to enhance cooperation among themselves, as well as collaboration with IGOs and Parties. NGOs were also invited to foster the participation of all stakeholders, particularly youth, women, civil society organizations and the media, in Article 6 implementation and to participate in the SBI Article 6 dialogue (Articles 25–27). Finally, all Parties were also requested to report on Article 6-related efforts in national communications and by other means, including the climate change information clearinghouse (CC:iNet) and on social media platforms (Articles 31–32). Table 2.9 summarizes the developments at COP17 (2011) and COP18 (2012).

Table 2.9 Summary of developments at COP17 (2011) and COP18 (2012)

- Created the Durban Forum on Capacity Building, a venue for in-depth discussions on capacity building based on various synthesized documents
- Concluded second review of Marrakech framework, decided to continue using the framework with improvements
- The Doha Work Programme on Article 6 of the Convention was established, based on the same six action areas as the amended New Delhi Work Programme, but with several new suggested implementation activities
- The Work Programme includes an annual dialogue, organized by the SBI, on Article 6
- The Work Programme is slated for review in 2020

The 2014 ministerial declaration and the capacity-building portal

At COP20, Ministers and Heads of State made a declaration on climate-related education and awareness-raising, in which they recognized the 'fundamental role' Article 6 priorities play in meeting Convention objectives (Decision 19/CP.20). Ministers therefore reaffirmed their commitment to promote and facilitate 'the development and implementation of education and public awareness programmes', public access to information, and public participation on climate change and its effects (Articles 1, 2). In the remainder of the declaration, Ministers did the following:

- Encouraged governments to develop strategies to incorporate climate change into curricula and climate change awareness objectives into design and implementation of national policies (Art. 3).
- Urged the Parties to give 'increased attention' to education, training, public awareness, public participation, and public access to information on climate change (Art. 4).
- Encouraged the Parties to participate in and benefit from the relevant work of intergovernmental panels and UN expert groups (Art. 5).
- Expressed resolve to 'cooperate and engage' through multilateral, bilateral and regional initiatives aiming to raise awareness and enhance education on climate change (Art. 6).
- Reaffirmed their commitment to implementation of the Doha Work Programme (Art. 7).

Because the UNFCCC secretariat was already, by COP20, obligated to 'collect, process, compile and disseminate' information important to the review process for the Marrakech capacity-building framework, it was a natural next step for the Parties to request that datasets from the secretariat's reports be uploaded to a web-based portal. The capacity-building portal, launched in June 2014, lets users search project activities, programmes, stakeholders, and funding sources and allocations, download results and visualize data in interactive charts and maps (Mead 2014). The portal is intended to facilitate identification of best practices, gaps and opportunities for cooperation, thereby enabling coordination of capacity-building support (UNFCCC website, Capacity-building portal page). Thus far, its catalogue includes data on 2684 activities, undertaken in developing countries from 1975 to 2019 (ibid.).

Plans are underway to expand the portal into a 'more ambitious', 'multifunctional' tool that includes information updated yearly, data extracted from a wider range of sources, such as national communications and needs assessments, information entered directly by intergovernmental organizations, and with functions that allow wider categories of results, more specific searches, and storage of data in a personal library (UNFCCC website, Capacity-building portal page). Table 2.10 summarizes developments at the Lima COP20 (2014).

Table 2.10 Summary of developments at the Lima COP20 (2014)

- The Ministerial Declaration on Article 6 reaffirmed the importance of education and awareness-raising and urged the Parties to do more to support relevant initiatives and incorporate relevant objectives into national policies
- COP20 was the first COP to make use of the UNFCCC's online capacity-building portal

Capacity building in the Paris Agreement

At COP21, in 2015, the question of what role the UN should play in capacity building – beyond repeatedly affirming its importance in enabling developing countries to implement the Convention and stressing the need to facilitate synergies – emerged as a topic of debate. Broadly, delegations from the Global North argued that their existing support for capacity building, mostly in the form of projects focused on fulfilling developing countries' specific Convention reporting requirements, and carried out by developed countries' development assistance agencies, should be allowed to continue without 'interference'. Negotiators from the Global South largely asserted that the Convention should play a more proactive role in coordinating and overseeing capacity building programmes and projects (Huq 2016b).

In both the COP21 decision text and the Paris Agreement, the Parties recognized the critical importance of capacity building. The decision establishes the Paris Committee on Capacity Building (PCCB) and the Capacity Building Initiative for Transparency (CBIT), while the devotion of Article 11 of the Agreement to capacity building and climate-related education definitively established them as crucial means to enhance climate action under the Paris Agreement (Dagnet et al. 2015).

Article 11 of the Paris Agreement lays out the goals, guiding principles and procedural obligations of all Parties to the Agreement with regard to capacity building. It states that capacity building under the Agreement should include a focus on 'countries with the least capacity, such as the least developed countries, and those that are particularly vulnerable to the adverse effects of climate change, such as small island developing States', and should 'be country-driven, ... foster country ownership of Parties, ... and should be an effective, iterative process that is participatory, cross-cutting, and gender-responsive' (Articles 11.1, 11.2). It also asserts that developed country Parties should support capacity building in developing countries and regularly communicate on relevant actions and measures (Articles 11.3, 11.4), while developing countries should also regularly communicate progress made on implementing capacity-building plans, policies, actions and measures (Art. 11.4). Finally, Article 11 tasks the first Conference of the Parties serving as the meeting of Parties to the Paris Agreement (CMA1) with consideration and adoption of a decision on initial institutional arrangements for capacity building (Art. 11.5).

The decision adopting the Paris Agreement established the Paris Committee on Capacity Building in order 'to address gaps and needs, both current and emerging, in implementing capacity-building in developing country Parties and further enhance capacity-building efforts, including with regard to coherence and coordination in capacity-building activities under the Convention' (Decision 1/CP.21, Art. 71). The decision also directed the SBI to organize annual in-session Paris Committee on Capacity Building meetings and to develop terms of reference for the Committee for adoption at COP22 (paras 75, 76). Most substantially, the decision instructed the PCCB to oversee a work plan for the period 2016–2020, including the following nine activities (Art. 73):

1 Assessing how to increase synergies and avoid duplication among Convention bodies, including through collaboration with institutions under and outside the Convention.
2 Identifying capacity gaps and needs and recommending ways to address them.
3 Promoting development and dissemination of tools and methodologies for capacity-building implementation.
4 Fostering global, regional, national and subnational cooperation.
5 Identifying and collecting good practices, challenges, experiences and lessons learned from capacity-building work by Convention bodies.
6 Exploring how developing countries can take ownership of building and maintaining capacity over time and space.
7 Identifying opportunities to strengthen capacity at the national, regional and subnational level.
8 Fostering dialogue, coordination, collaboration and coherence among relevant processes and initiatives under the Convention, including through exchanging information about activities and strategies of Convention bodies.
9 Providing guidance to the secretariat on the maintenance and further development of the web-based capacity-building portal.

The Paris decision text also establishes the *Capacity Building Initiative for Transparency* (CBIT), discussed at length in Chapter 8, 'in order to build institutional and technical capacity, both pre- and post-2020' to support developing countries' efforts to meet the 'enhanced transparency requirements' laid out in Article 13 of the Agreement (Decision 1/CP.21, Art. 84). The CBIT will aim to strengthen national institutions for transparency-related activities in line with national priorities, provide relevant tools, training and assistance for meeting Article 13 provisions, and assist in the improvement of transparency over time (Art. 85). Accordingly, Article 13.15 of the Agreement stipulates that 'support shall also be provided for the building of transparency-related capacity of developing country Parties on a continuous basis'.

The provision of finance for capacity building under Article 11.3 of the Paris Agreement is a recommendation ('Developed country Parties *should* enhance support for capacity-building actions in developing country Parties', emphasis added), while the provision of support for the CBIT under Article 13 is obligatory ('Support *shall* be provided to developing countries for the implementation of this Article', 'Support *shall* also be provided for the building of transparency-related capacity of developing country Parties on a continuous basis', emphasis added). It is noteworthy, however, that Article 13 does not specify any group of countries that shall provide support. These different formulations and omissions provide leeway for interpretations by the Parties in negotiations. The 'should'/'shall' discrepancy also suggests that powerful Parties are perhaps more interested in ensuring capacity building for transparency than addressing overall capacity gaps. Table 2.11 summarizes developments at the Paris COP21 (2015).

Back to Marrakech: specifying the Paris Committee on Capacity Building

At COP22, in Marrakech, Morocco, in late 2016, the Parties began to elaborate the necessary specifics for implementation of the Paris Agreement, including decisions on the structure and future work of the Paris Committee on Capacity Building (PCCB). The COP also completed the third review of the Marrakech framework for capacity building in developing countries.

At SBI 44, in May 2016, the Parties developed terms of reference for the Committee, which were then adopted at COP22:

- The Paris Committee on Capacity Building will be composed of 12 members, including two from each of the five UN regional groups, one from the LDCs, and one from the SIDS (Decision/CP.22, Annex, Art. 2).

Table 2.11 Summary of developments at the Paris COP21 (2015)

- Article 11 established capacity building as an important means of implementation under the Paris Agreement
- The Paris Committee on Capacity Building was created to enhance coordination of capacity building activities under the Convention and thereby improve efforts to address capacity-building needs
- The Capacity Building Initiative for Transparency was established to aid developing countries in implementing the transparency requirements of Article 13
- Financial support by developed countries is recommended by Article 11, while financial support is required by Article 13 but its source is unspecified

- Members will be nominated by their respective groups and elected by the COP, and groups are encouraged to make nominations with a view towards 'achieving an appropriate balance of experts relevant to the aims of the Committee' and achieving a gender balance (Art. 3).
- Members will serve for two years for a maximum of two consecutive terms. Six seats will be up for election every year (Art. 4).
- The Paris Committee on Capacity Building will elect two co-chairs annually from among its members, to serve for a term of one year (Art. 6).
- Six representatives from Convention bodies and operating entities of the Financial Mechanism will be invited to participate in meetings for a term of one year (Articles 3, 4).
- The PCCB will meet annually, during SBI sessions (Art. 9).
- The PCCB will decide on its annual focus area, related to enhanced technical exchange on capacity building (Art. 10).
- The PCCB may invite other Convention bodies and operating entities of the Financial Mechanism to identify representatives for further collaboration (Art. 12).
- The PCCB may engage with relevant institutions, organizations, frameworks, networks and centres outside the Convention, at multiple levels, and draw upon their expertise (Art. 13).
- PCCB meetings will be open to admitted observer organizations, except where otherwise decided by the committee, 'with a view to encouraging a balanced regional representation of observers' (Art. 14).

The Parties also reaffirmed the objective of the Paris Committee on Capacity Building as it was described in the Paris Agreement and recalled that a review of the Paris Committee on Capacity Building, including its progress and need for extension and enhancement, will proceed at COP25 (Decision/CP.22, Articles 2, 3). Finally, Parties requested the SBI organize the first meeting of the Paris Committee on Capacity Building in conjunction with SBI 46, to take place in May 2017 (Art. 6).

In the third comprehensive review of implementation of the Marrakech capacity building framework in developing countries, the Parties reaffirmed that the framework is still relevant, and that capacity building is an 'integral component ... enabling developing country Parties to implement the Convention and the Paris Agreement' (Decision/CP.22, Preamble, para. 1). Within the review, Parties invited the Paris Committee on Capacity Building, in managing its 2016–2020 work plan, to take into consideration:

- cross-cutting issues: gender responsiveness, human rights, indigenous knowledge;
- outcomes of the third comprehensive review;
- previous work undertaken on capacity-building indicators;
- ways of enhancing reporting on capacity-building activities.

And to promote and explore:

- linkages with other Convention bodies;
- synergies for enhanced collaboration with institutions outside the Convention engaged in capacity-building activities (Art. 4).

The review decision also invites Parties to undertake a number of actions. These include considering how to enhance reporting on the impacts of capacity-building activities and how lessons learned are fed back into relevant processes (Art. 3) and 'foster[ing] networking and enhanc[ing] their collaboration with academia and research centres, with a view to promoting individual, institutional, and systemic capacity-building through education, training, and public awareness' (Art. 5). Developed country Parties are also invited to 'enhance support for capacity-building actions in developing country Parties' (Art. 7).

Finally, the decision invites 'relevant intergovernmental and non-governmental organizations, as well as the private sector, academia, and other stakeholders, to continue incorporating into their work programmes' capacity-building needs, and 'also invites United Nations agencies, multilateral organizations and ... observer organizations engaged in providing capacity-building support to ... provide information ... to be uploaded in the capacity-building portal' (paras 8, 9).

At that, the Parties concluded the third comprehensive review of the Marrakech framework for capacity building in developing countries and decided to initiate the fourth at SBI 50 (June 2019), to be completed at COP25 (November 2019). Table 2.12 summarizes developments at COP22 (2016).

Continued focus on capacity building in 2017

As was the case with COP22, COP23 was devoted to developing the rulebook for implementation of the Paris Agreement. The Subsidiary Body for Implementation (SBI) also considered the annual technical progress report of the Paris Committee on Capacity Building (PCCB), which covered the work the PCCB undertook between its first meeting, held in May 2017, and August 2017, including information on PCCB membership, rules of

Table 2.12 Summary of developments at COP22 (2016)

• Terms of reference for the Paris Committee on Capacity Building, including membership details, were adopted
• The Marrakech framework for capacity building in developing countries was reviewed and deemed still relevant. Parties and the Paris Committee on Capacity Building were invited to consider how to improve reporting and enhance collaboration, and developed country Parties were also invited to enhance support for capacity building in developing countries

procedure, and the PCCB's rolling workplan for 2017-2019. Parties appreciated the progress presented in the annual report, as well as its analysis of capacity-building needs and gaps in the context of Nationally Determined Contribution implementation. The SBI and PCCB will continue their work in 2018 and, based upon their recommendations, Parties at COP24 will finalize the institutional arrangements on capacity-building to support the Paris Agreement under Article 11, paragraph 5 ("Capacity-building activities shall be enhanced through appropriate institutional arrangements to support the implementation of this Agreement").

Also during COP23, the International Centre for Climate Change and Development co-hosted with GIZ GmbH (the German Corporation for International Cooperation) the first-ever "Capacity Building Day" at a COP. The event was attended by the LDC Group Chair, PCCB delegates, and other UNFCCC delegates, and included presentations on two nascent initiatives for building alliances among universities for climate action, the LDC University Consortium for Climate Change and the University Network for Climate Capacity.

Conclusion: politics in climate negotiations on capacity building

The long history just recounted shows how layer upon layer of expectation and tracking and sharing of capacity building efforts has piled up over 25 years of negotiations. Capacity building under the UNFCCC has been called a 'motherhood-and-apple-pie issue', as it has proven to be a generally uncontroversial item upon which the Parties can reliably agree in negotiations. The lack of contention that typically surrounds capacity building has historically convinced industrialized countries to send their most junior delegates to capacity-building negotiation meetings (Huq 2016b).

Capacity building's uncontroversial nature is largely a result of its broad construction within the text of the Convention and within development theory, which allows a wide variety of efforts to be classified as 'capacity building' and has facilitated complex intertwinement of capacity building with other Convention objectives and implementation programmes. First, the founding text of the Convention expounds an understanding that finance for capacity building is necessary to enable developing countries to undertake mitigation and adaptation efforts, giving capacity building a basis that was essentially unassailable in subsequent decisions aiming to achieve the Convention's objectives. The question of whether capacity building is an important means to the ends of mitigating and adapting to climate change was never up for debate; instead, negotiations focused on modalities for providing finance and delivering both hard and soft technologies. Second, capacity building, as incorporated in the text of the Convention, encompasses many types of efforts, from those with ends as concrete as constructing a solar power plant to those with ends as intangible as improving public understanding of climate change. Because such a wide range of efforts could be called capacity building and technology transfer, early COP decisions aiming to enhance developing countries' capacities removed little latitude from project funders and therefore evoked little discord from among developed country Parties. And when it comes to capacity-building efforts on climate, the world always needed more. Much more.

The Marrakech frameworks have guided capacity-building activities under the Convention since their establishment at COP7 in 2001, enduring the creation of various further work programmes and forums on capacity building. But, following 15 years of activities undertaken according to the needs and priorities of developing countries as identified in the frameworks, many of the problems that motivated the creation of the frameworks 'remain the core concerns communicated annually by Parties' (Dagnet et al. 2015). Because no Convention body was allotted a responsibility to coordinate or evaluate capacity-building actions performed under the framework, it is difficult now to take stock of progress made versus needs remaining on each priority area. It is evident, however, that due to the lack of central oversight and coordination, priority areas have been unevenly addressed: for example, broadly speaking, institutional capacity building (i.e. strengthening or establishment of national climate change secretariats or national focal points) does not appear to have been pursued to the same extent as the objective of building countries' capacities to manage climate and emissions data and use it to create national GHG inventories and national communications (ibid.).

Although this history shows that capacity building was consistently given attention at COPs and intersessional meetings preceding Paris, decisions were largely redundant, repeatedly urging cooperation and synergies, inviting participation by relevant stakeholders, and calling for enhanced support for capacity building. The Paris COP was the first to produce an outcome tasking a committee with overseeing capacity-building objectives, and, accordingly, hosted the most contentious capacity-building negotiations in recent history.

Leading up to Paris, developing country representatives urged the improvement of coordination and coherence of capacity building under the UNFCCC. On behalf of the LDC group, Angola stated in a submission to the secretariat ahead of the Fourth Durban Forum on Capacity Building that the delivery of capacity-building support on an ad-hoc basis, attached to individual, time-bound projects, had not created a sustainable structure to build capacity over the long term, and that treatment of capacity building as a cross-cutting issue that should be universally pursued would remain flawed as long as no entity was charged with coordinating and overseeing capacity building activities (ibid.). Similarly, on behalf of the Africa Group, Sudan underscored the need for improved coordination of support for capacity building among donors, with recipient nations' priorities, and with emerging capacity gaps in developing countries (ibid.). Conversely, developed countries maintained the lack of specificity characteristic of their provision of capacity-building support in the run-up to the Paris COP, and representatives of the Global North seemed 'quite content with their current approach' of ad-hoc, short-term, uncoordinated capacity building efforts (Huq 2016b). During the Paris COP, predictably, developing country delegates argued that the current approach to capacity building was inefficient, costly, frequently unsustainable and in need of improvement, while developed country representatives pushed back against proposed changes (ibid.).

Although the Paris negotiations concluded with relatively substantial developments on capacity building – including the creation of the Paris Committee on Capacity Building to oversee and coordinate capacity-building efforts and the establishment of the CBIT to support transparency in developing countries' climate action – politics, especially in terms of clashes between the Global North and the Global South, precluded a number of further developments. For example, no guidance was provided in Paris as to the form (i.e. terms of reference or membership) of the Paris Committee on Capacity Building due to the Parties' disagreements on the proper reach of the Committee's oversight and coordination powers. However, developments at COP22 and COP23 appeared to bode well for the future of capacity building under the Convention, as progress was made on both the CBIT and Paris Committee on Capacity Building, including $55.3 million in pledges to the CBIT from 11 developed countries and the adoption of Terms of Reference and a rolling 2017-2019 workplan for the Paris Committee on Capacity Building (GEF 2016a). The potential of the Convention to overcome the developed-developing world divide on whether to define capacity building, systematically monitor and evaluate capacity-building efforts, and coordinate capacity-building strategies will be determined largely by the work of the Paris Committee on Capacity Building over the coming years, beginning with its first meeting in May 2017. Whether the Capacity Building Initiative on Transparency will be reliably and consistently funded is another key determination to be made in the near future.

References

Climate Change Information Network. UNFCCC. Available at: http://unfccc.int/cc_inet/cc_inet/items/3514.php (accessed 7 October 2017).

Dagnet, Y., Northrop, E., & Tirpak, D. (2015). *How to strengthen institutional architecture for capacity building to support the post-2020 climate regime.* Washington, DC: World Resources Institute.

GEF. (2016a) Joint statement on the donors' pledge of $55.3M to the Capacity-building Initiative for Transparency. Washington, DC: Global Environment Facility. Available at: www.thegef.org/sites/default/files/web-documents/CBIT-donor-statement-COP22.pdf (accessed 29 December 2016).

GEF. (2016b). *50th GEF council meeting, June 07–09, 2016: Programming directions for the Capacity-Building Initiative for Transparency*. Washington, DC: Global Environment Facility. Available at: www.thegef.org/gef/sites/thegef.org/files/documents/EN_GEF.C.50.06_CBIT_Programming_Directions_0.pdf (accessed 29 October 2016).

Huq, S. (2016a). *Why universities, not consultants, should benefit from climate funds*. Available at: www.climatechangenews.com/2016/05/17/why-universities-not-consultants-should-benefit-from-climate-funds/

Huq, S. (2016b). *Stop sending climate consultants to poor countries – invest in universities instead*. Available at: http://theconversation.com/stop-sending-climate-consultants-to-poor-countries-invest-in-universities-instead-65135

Mead, L. (2014) UNFCCC launches capacity-building portal. IISD SDG Knowledge Hub. Available at: http://sdg.iisd.org/news/unfccc-launches-capacity-building-portal/?rdr=climate-l.iisd.org (accessed 7 November 2016).

UNFCCC. (United Nations Framework Convention on Climate Change). (n.d.). A guide to the UNFCCC and its processes: Capacity-building. Available at: http://bigpicture.unfccc.int/ (accessed 2 November 2016).

UNFCCC. (1999). Annex to Decision 10/CP.5. List of capacity-building needs of developing country parties. Available at: http://unfccc.int/resource/docs/cop5/06a01.pdf

UNFCCC. (2001a). Decision 2/CP.7, Annex. Framework for Capacity Building in Developing Countries. Available at: http://unfccc.int/resource/docs/cop7/13a01.pdf

UNFCCC. (2001b). Decision 2/CP.7. Capacity building in developing countries (non-Annex I Parties). Available at: http://unfccc.int/resource/docs/cop7/13a01.pdf#page=5

UNFCCC. (2002). Decision 11/CP.8. New Delhi Work Programme on Article 6 of the Convention. Available at: http://unfccc.int/resource/docs/cop8/07a01.pdf#page=23

UNFCCC. (2004). Decision 2/CP.10. Capacity-building for developing countries (non-Annex I Parties). Available at: http://unfccc.int/resource/docs/cop10/10a01.pdf#page=7

UNFCCC. (2005). Decision 29/CMP.1. Capacity-building relating to the implementation of the Kyoto Protocol in developing countries. Available at: http://unfccc.int/resource/docs/2005/cmp1/eng/08a04.pdf#page=5

UNFCCC. (2007). Decision 9/CP.13. Amended New Delhi Work Programme on Article 6 of the Convention. Available at: http://unfccc.int/resource/docs/2007/cop13/eng/06a01.pdf#page=37

UNFCCC. (2009a). Decision 2/CP.15. Copenhagen Accord. Available at: http://unfccc.int/resource/docs/2009/cop15/eng/11a01.pdf#page=5

UNFCCC. (2009b). Decision 11/CP.15. Administrative, financial and institutional matters. Available at: http://unfccc.int/resource/docs/2009/cop15/eng/11a01.pdf#page=29

UNFCCC. (2010). Decision 1/CP.16. The Cancun Agreements: Outcome of the work of the Ad hoc Working Group on Long-Term Cooperative Action under the Convention. Available at: http://unfccc.int/resource/docs/2010/cop16/eng/07a01.pdf#page=2

UNFCCC. (2011a). Cancun Adaptation Framework. Available at: http://unfccc.int/resource/docs/2010/cop16/eng/07a01.pdf#page=4

UNFCCC. (2011b). Decision 13/CP.17. Capacity-building under the Convention. Available at: https://unfccc.int/files/cooperation_and_support/capacity_building/application/pdf/decision_13cp17.pdf

UNFCCC. (2011c). FCCC/CP/2011/9/Add.1. Decision 2/CP.17. Outcome of the work of the Ad hoc Working Group on Long-Term Cooperative Action under the Convention. Available at: http://unfccc.int/resource/docs/2011/cop17/eng/09a01.pdf#page=4

UNFCCC. (2012). Decision 15/CP.18. Doha Work Programme on Article 6 of the Convention. Available at: http://unfccc.int/resource/docs/2012/cop18/eng/08a02.pdf#page=17

UNFCCC. (2014). Decision 19/CP.20. The Lima Ministerial Declaration on Education and Awareness-raising. Available at: http://unfccc.int/resource/docs/2014/cop20/eng/10a03.pdf#page=37

UNFCCC. (2015). Decision 1/CP.21. Adoption of the Paris Agreement. http://unfccc.int/resource/docs/2015/cop21/eng/10a01.pdf

UNFCCC. (2016a). Decision-/CP.22. Paris Committee on Capacity-building. Available at: http://unfccc.int/files/meetings/marrakech_nov_2016/application/pdf/auv_cop22_i13_Paris Committee on Capacity Building.pdf

UNFCCC. (2016b). Decision-/CP.22. Third comprehensive review of the implementation of the framework for capacity-building in developing countries under the Kyoto Protocol. Available at: http://unfccc.int/files/meetings/marrakech_nov_2016/application/pdf/cmp12_i10_3rd_comprehensive_review_for_kp_cb.pdf

UNFCCC, Capacity-building portal. Available at: http://unfccc.int/capacity building/core/activities.html (accessed 3 November 2016).

3 Has it worked elsewhere?

Capacity-building efforts in development and environmental regimes

Introduction

Capacity building as an element of development cooperation has been going on in many different names and forms for more than half a century. These efforts have been pursued under a diverse set of global regimes that have evolved during the last few decades. One category of such regimes relates to development issues, such as economic development, poverty reduction and sustainable development, which involves different kinds of assistance from bilateral and multilateral agencies. Trade and regional economic integration represents a second set of relevant regimes with distinct objectives and regulatory instruments. A third category of international regimes pertains to the protection of the environment and natural resources, and a fourth is the international human rights framework, with its array of covenants and declarations. A fifth and final category applies to security, cooperation, and humanitarian affairs. While there are thematic overlaps among all these regimes, each category represents specific constituencies and possesses its own normative and conceptual frameworks, procedures, institutions and approaches. Understanding and reconciling these multiple types of regimes and their relevance to national and international development present considerable challenges.

Research on international institutions has identified capacity as one of three conditions for their effectiveness, the other two being sufficient concern and solutions to the identified problems (Hass et al. 1993) Under each regime category, different kinds of capacity-building activities have been undertaken with donor support in the developing countries. There are both generic and regime-specific capacities which have been addressed through technical cooperation programmes of the rich industrial countries. This chapter is devoted to reviewing capacity-building initiatives under different regimes, including the World Trade Organization (WTO), the Regional Seas Programme, the human rights regime, the disaster risk reduction regime, and the Montreal Protocol regime on ozone-depleting substances. The purpose of this exercise is to look at similarities and differences in approaches to capacity building and to learn lessons for charting a realistic course for capacity building to address climate change.

International trade and capacity building

Developments in capacity building under the WTO

International trade and even globalization are not new phenomena. However, the globalization that began in the 1990s bears extra significance for several reasons: First, with the collapse of Soviet communism, the world was transformed largely into one market system. Second, developing countries with their rapid economic growth have recently been occupying a significant position in the global economic and trade systems. Third, improvements in technology of production, transport and communications have boosted international trade of goods and services increasingly since the 1950s. The World Trade Organization (WTO), established in the mid-1990s,replaced the General Agreement of Tariff and Trade (GATT) that took shape after the Second World War, and promoted this expansion through the liberalization of global trade. The technical aspects of trade policy making in the context of the WTO and other forums have become increasingly complicated, with 'new' issues such as services, intellectual property, technical barriers to trade (TBTs), e-commerce, etc. (OECD 2006a).

Thus, the WTO began helping the developing countries and LDCs to take advantage of a globalized economic and trading system. The WTO has done so through binding commitments to different trade regimes, select flexibilities, and well-crafted technical assistance for building institutional infrastructure in developing countries (WTO 2014). The Committee on Trade and Development (CTD), as the focal point on development issues in the WTO, plays an important role by considering issues raised by developing countries and specific groups, such as the LDCs and other small economies. It allows flexibilities in commitments, supports preferential tariff treatment and particularly promotes the trade-related technical assistance and capacity-building programmes.

One of the mandated tasks of the WTO is to assist the developing countries in trade policy issues through technical cooperation (TC) and training programmes. The Ministerial Declaration in Doha (WTO 2001: para. 2) declared:

> The majority of WTO Members are developing countries. We seek to place their needs and interests at the heart of the Work Programme adopted in this Declaration. Recalling the Preamble to the Marrakesh Agreement, we shall continue to make positive efforts designed to ensure that developing countries, and especially the least-developed among them, secure a share in the growth of world trade commensurate

> with the needs of their economic development. In this context, enhanced market access, balanced rules, and well-targeted, sustainably financed technical assistance and capacity-building programmes have important roles to play.

Thus, the Doha Declaration stresses the important role of sustainably financed technical assistance and capacity-building programmes. The Declaration dedicates a whole section to these issues, and reiterates the commitment of its membership to help the weakest among them participate effectively in the multilateral trade system (MTS).

The Declaration further stipulates:

> We have established firm commitments on technical cooperation and capacity building in various paragraphs in this Ministerial Declaration. We reaffirm these specific commitments contained in paragraphs 16, 21, 24, 26, 27, 33, 38–40, 42 and 43, and also reaffirm the understanding in paragraph 2 on the important role of sustainably financed technical assistance and capacity-building programmes.
>
> (para. 41)

In 1997, a high-level meeting on trade initiatives and technical assistance for the LDCs resulted in an 'integrated framework' (IF) involving six intergovernmental agencies, to help LDCs increase their capacity to trade, and preferential market access agreements. The WTO CTD, assisted by a Sub-Committee on LDCs, looks at their special needs. Its responsibility includes implementation of the agreements, technical cooperation issues, and the increased participation of developing countries in the multilateral trade system. After several years of the integrated framework's operation, it was evident that gaps in capacity building and efforts to mainstream trade into development strategies persist (EIF 2006). Therefore, an Enhanced Integrated Framework (EIF) became operational from the beginning of 2009, with additional financial resources and institutional support.

Elements of the integrated capacity-building framework

It is obvious from the above statements that there was a firm and unequivocal commitment on the part of the global community to promote capacity-building activities targeted at bringing the developing countries into the fold of the global trade regime. Accordingly, trade-related technical assistance activities have encompassed the following elements:

1 Institution-building and strengthening to handle trade policy issues, i.e., assistance to LDCs in acceding to the WTO; enhancing capacities to make and implement trade policy consistent with the WTO obligations; seeking effective coordination among government departments;

building a 'core capacity' to deal with trade issues within a lead Ministry and the development of research and analytical capacity on trade issues; strengthening capacity to participate in the multilateral trade system, including the implementation and application of obligations and commitments; accessing relevant information for negotiations on trade issues.
2 Strengthening export capabilities, that is, strengthening the policy environment for trade liberalization; improving competitiveness of enterprises; increasing investments in productive sectors; removing bottlenecks to increased production of tradable goods and services, including through development of relevant infrastructure; helping the LDCs exploit new trade opportunities.
3 Strengthening trade support services, e.g. access to trade finance; support for access to business information, use of information technology, development of new products and trade diversification, advice on standards, packaging, quality control, marketing and distribution channels; commercial representation; functioning of trade promotion organizations; improved purchasing and supply management; promotion of trade in services, etc.
4 Strengthening trade facilitation capabilities, e.g. modernization and reform of customs services and other government agencies participating in trade transactions, simplifying export and import procedures.
5 Professional training and human resource development, a large component in each of the above areas.
6 Assistance in the creation of a supportive trade-related regulatory and policy framework that will encourage trade and investment.

At the national level, an elaborate institutional architecture was put in place to oversee the implementation of the above initiatives. This includes an Enhanced Integrated Framework (EIF) National Steering Committee, a National Focal Point, a national level EIF Facilitator (usually a donor representative) and a Local Project Appraisal Committee, comprised of government, donor, private sector and civil society representatives.

Approaches and tools for capacity building

The steps and procedures under the Integrated Framework are as follows: needs assessment, inter-agency coordination, implementation and follow-up (WTO 1997). The following tools have been used so far for capacity building:

- Offering of trade courses: on average three trade policy courses each year held in Geneva for government officials. A Progressive Learning Strategy (PLS) under the WTO allows participants to register for training at different levels (introductory, intermediate and advanced) depending on their familiarity with the subject. They can also choose a generalist or a specialist path, according to their professional needs (WTO 2014). In recent years a few courses have been offered, such as Trade Services Statistics, WTO Food Security Agreements, Enhancing LDC Participation in the multilateral trade system (MTS), etc.

- National and regional seminars/workshops are held regularly in all regions of the world with a special focus on African countries. The participants include parliamentarians, government officials, businessmen, NGOs and civil society members.
- The WTO has set up reference centres in over 100 trade ministries and regional organizations in capitals of developing countries and LDCs. These centres provide computers and internet access to enable ministry officials to keep abreast of events in the WTO through online access to the WTO's huge database of official documents and other materials. Attempts are also being made to help countries that do not have permanent representatives in Geneva.

Funding for trade capacity building

Since the inception of the WTO, under the banner of trade-related technical assistance (TRTA), a host of funds in different names have been initiated to assist developing countries, LDCs and other small economies in building capacity for multilateral trading. Box 3.1 briefly presents the list with their main missions. Compared to capacity-building programmes under some other regimes, the funding and structures for WTO capacity building appear relatively substantial. Still, even after repeated pledges at Ministerial meetings, actual financial commitments have remained small, well below what the rhetoric suggests. The setting-up of the 'Doha Development Agenda Global Trust Fund' (DDAGTF) as the umbrella financing mechanism and various other efforts by donors, such as the European Commission (EC), have only slightly improved the situation.

There are incentives for OECD members to provide capacity building to developing countries and LDCs so that they can speed up implementation of the Trade Facilitation Agreement (TFA) (WTO 2015). Developed countries as the main exporter of manufactured goods, services and technology benefit immensely from a liberalized trade where all developing countries mainstream trading as an engine of growth. Furthermore, inefficient trade procedures create losses that affect all parties involved in a multilateral trade system. By providing assistance and support for capacity building to developing countries and LDCs, developed countries also reduce or eliminate the losses faced by their firms (ibid.).

With elaborate institutional and funding support, trade capacity building under WTO appears to have been quite successful. Many of the factors contributing to its success are interrelated, and in several cases they are mutually supportive. In addition, different trade facilitation measures often

Box 3.1 Trade-related technical assistance funds for capacity building

1 *The Doha Development Agenda Global Trust Fund* (DDAGTF) is the flagship WTO fund providing support to a wide range of trade-related technical assistance (TA) activities. Established in 2002, it receives extra budgetary contributions from WTO Members to finance the implementation of the annual TA plans. This Global Trust Fund operates against periodic benchmarks supervised by the Committee on Budget, Finance and Administration and the Committee on Trade and Development (CTD). Total cost of the TA Plans is budgeted at around CHF20 million annually, of which some CHF15 million is covered from Trust Funds and the remaining from the WTO's regular budget. The WTO's Biennial Technical Assistance and Training Plans indicate how the assistance is provided.

2 *Trade Facilitation National Needs Assessments:* Provides resources to allow participants to establish preliminary trade facilitation needs and priorities.

3 *Trade Facilitation Negotiating Group:* Provides resources for officials from developing countries to participate in Geneva-based meetings of the Negotiating Group on Trade Facilitation.

4 *Internship Programme – Selected Missions:* Provides resources for interns from developing countries to gain experience with the MTS by working at the Geneva missions of their own countries.

5 *The Netherlands Trainee Programme:* Provides resources for interns from developing countries to gain experience with the multilateral trade system (MTS) by working at the WTO Secretariat.

6 *China LDCs & Accessions Programme:* Provides resources for LDC interns to work on accession issues at the WTO Secretariat, for LDC officials to participate in Geneva-based meetings, and to organize roundtables on accession issues.

7 *Standards and Trade Development Facility:* The STDF is a joint initiative in capacity building and technical cooperation aimed at raising awareness of the importance of sanitary and phytosanitary (SPS) issues, increasing coordination in the provisions of SPS-related assistance, and mobilizing resources to enhance developing countries' capacity to meet SPS standards.

8 *The Aid for Trade* initiative is supported by a broad range of intergovernmental organizations and is aimed at helping developing countries mainstream trade into their development strategies and mobilize donor support for capacity building and trade-related infrastructure. The initiative was launched at the WTO's Ministerial in Hong Kong in 2005. Since 2007, the WTO has also held Global Reviews of Aid for Trade every two years. Since 2009 this EIF Trust Fund has had two tiers: Tier 1 helps to incorporate trade into national development plans and to translate trade priorities into bankable projects for broader Aid for Trade funding; Tier 2 provides bridging funding to 'jump start' activities in project preparation, feasibility studies and funding of seed projects.

9 *Trade Facilitation Agreement Facility (TFAF):* In December 2013, WTO members concluded negotiations on a Trade Facilitation Agreement at the Bali Ministerial Conference, and Section II of the Agreement established a link between the obligations of developing countries and their implementation capacity. Under this scheme, developing countries are allowed to determine their own technical assistance needs and implementation schedule. The TFAF Trust Fund was launched in 2014 to ensure that developing countries and LDCs receive the assistance they need to implement and benefit from the Trade Facilitation Agreement (TFA), whose main purpose is to reform the customs and tariff administration policies in developing countries. The TFA Facility supports LDCs and other developing countries in assessing their specific needs and identifying possible development partners to help them meet those needs, ensuring the best possible flow of information between donors and recipients through the creation of an information-sharing platform for demand and supply of trade facilitation-related technical assistance. The TFAF expanded the existing WTO trade-related technical assistance programme for parliamentarians to have a greater focus on trade facilitation. In 2014, trade facilitation workshops for parliamentarians were conducted for African countries, the Eastern African Community, ASEAN countries, all Latin American countries, and the Pacific Islands.

Source: WTO (2014, 2015, 2016)

involve different types of success factors. Keeping this in mind, the factors can be grouped into six broad categories: (1) national ownership; (2) stakeholder participation; (3) financial, material and human resources; (4) a sequencing approach; (5) transparency and monitoring; and (6) other factors (WTO 2014).

Let us focus on financial, human and material resources. In 95 case studies, it was found that funding mechanisms, whether financial support was domestic, external, or a combination of both, were important. In particular, a relatively higher number of case studies on trade facilitation projects in LDCs underscores the key role played by adequate, predictable and reliable donor funding (WTO 2015).

Adequate human resources and organizational management, mentioned in 61 case studies, are also reported as critical elements in enhancing the quality and integrity of staff with respect to the trade facilitation. Besides training and professional development, the remuneration, incentives, promotion, rotation and relocation offered to staff may have to be considered to ensure that they internalize the objectives of the trade facilitation reform and accept their new role and responsibilities (Vinod 2006).

Finally, some observations can be made as a critique of the WTO trade-related technical assistance (TRTA) system and process. These issues have been discussed in detail (Lecomte 2001). One common shortcoming of trade capacity development projects is that the TRTA may end up being biased in favour of donors rather than recipients. Unlike technical assistance in other sectors, TRTA involves conflicts of interest between donors as trading powers and new traders. Several trade officials argued that awareness and leadership are the best antidotes to biased TRTA.

Second, evidence suggests that donor coordination in TRTA may be even more difficult than in other aid areas(OECD 2006b). Constrained donor capacity in this sector may account for a lack of coordination. One report posits that the WTO's Institute for Training and Technical Cooperation (ITTC) strengthened its coordinating role for trade capacity building (CB-EIF-WB).

Third, TRTA projects often address specific aspects of trade policy process, with very few taking a comprehensive approach. One programme – the ITC/WTO/UNCTAD Joint Integrated Technical Assistance Programme (JITAP) – seems to be an exception – its approach, aimed at promoting an inclusive trade policy process encompassing a wide range of stakeholders, appears a unique and inspiring feature (UNCTAD & UNDP 2006).

Fourth, the results of TRTA are likely to be less tangible than in other sectors such as health or education. This may be a handicap, as donor agencies often work under pressure to spend project funds quickly and achieve visible results. But investing in the institutional capacity of public and private actors to engage in trade-related issues cannot produce visible and quantifiable results in the short run. This calls for new approaches to TRTA project assessment. However, during the last few years, results-based management has gained ground in the design, management, delivery and evaluation of capacity building programmes (World Bank 2004a).

Finally, donors may face a 'legitimacy problem' with their domestic constituencies, as trade impact on poverty alleviation is not often clear. So it is crucial to show the positive effects of trade on the poor, at least in the medium to long term.

The Regional Seas Programme (RSP)

Initiatives under the RSP

Rapid economic and population growth are wreaking havoc on the world's oceans and seas. Almost half of the world's population lives close to coastal areas. Over 80 per cent of ocean pollution emanates from land-based activities (GESAMP 1991). Key challenges in protecting regional seas are land-based pollution, marine litter from ships and boats, oil spills, bleaching of coral reefs, extinction and degradation of marine species, etc.

In order to address these challenges, UNEP initiated the first Regional Seas Programme (RSP) in the mid-1970s. RSPs now cover 18 regions of the world, making them one of the most globally comprehensive initiatives for the protection of marine and coastal environments. Table 3.1 presents the main features of these programmes (UNEP 2016). Some of them have been initiated by UNEP and others by independent regional initiatives.

Table 3.1 The Regional Seas Programme

Type of Regional Seas Programme	*Main features*	*Regions concerned*
UNEP administered Regional Seas Programme	Secretariat, administration of the Trust Fund and financial and administrative services provided by UNEP.	Caspian Sea East Asian Seas Mediterranean North-West Pacific Western, Central and Southern Africa Western Indian Ocean Wider Caribbean
Associated Regional Seas Programme	Secretariat not provided by UNEP. Financial and budgetary services managed by the programme itself or hosting regional organizations. UNEP support/collaboration were or are provided.	Black Sea North-East Pacific Pacific Red Sea and Gulf of Aden ROPME Sea South Asian Seas South-East Pacific
Independent Regional Seas Programme	Regional framework not established under the auspices of UNEP. Invited to participate in regional seas coordination activities of UNEP through the global meetings of the RSP. UNEP is also invited to participate in their respective meetings.	Antarctic Arctic Baltic Sea North-East Atlantic

Source: UNEP (2016), Table 3, p. 25.

Usually,RSPs function through regional action plans (RAPs), which are adopted by member governments in order to establish a comprehensive strategy and framework to protect the environment and promote sustainable development. The RAPs outline the strategy, based on region-specific environmental challenges and political and socio-economic conditions.

Fourteen of the RSPs also have adopted legally-binding conventions which express the commitment and political will of governments to tackle their common environmental issues through joint coordinated activities. Most conventions have added protocols for addressing specific issues such as protected areas, land-based pollution, or marine biodiversity.

As noted in the Global Strategic Review of the Regional Seas Programme (Ehler 2006):

> The RSP, its conventions and protocols, and action plans have provided a forum for equitable participation by Member States in management processes of major seas of the world. It has promoted the idea

> of a 'shared sea,' and has helped place marine and coastal management issues on the political agenda and supported the adoption of environmental laws and regulations. For some Member States in some regions, the RSP is the only entry point for environmental concerns. It has encouraged and provided assistance for capacity building for marine and coastal management.

Although many RSPs have made a positive difference, many have failed to solve the problems they were designed to address (ibid.). Several factors currently limit their effectiveness in tackling marine and coastal challenges. The implementation of regional agreements is far from systematic and comprehensive. The most glaring example is the disconnect between the number of regional agreements aimed at preventing land-based pollution and the persistence, and even worsening, of the problem in some cases (UNGA 2011). Many reasons, often cumulative, can explain this situation, including lack of political will, political instability in some states or weak enforcement mechanisms. The First Inter-Regional Programme Consultation identified 'the lack of necessary interaction with the fisheries sector and other socio-economic sectors' as one of the 'most fundamental problems hampering the implementation of the respective Regional Seas programmes' (UNEP 2001). Some generic challenges in protecting the regional seas include:

1. Absence of adequate scientific and managerial expertise in most cases.
2. Lack of scientific data and of access to technologies necessary to compile such data, relating to physical and biological patterns.
3. Lack of an integrated approach involving all sectors and stakeholders.
4. Inadequate financial resources to address the challenges.
5. Finally, lack of political stability in some cases, lack of political will and absence of adequate policy and institutional frameworks in many coastal countries.

These challenges can be addressed with appropriate measures, which are considered in the RSPs in many different forms. Such measures are:

1. Human resource development on a sustainable basis through setting up of degree programmes in disciplines, such as oceanography, coastal zone management, and marine conservation biology, as well as regular short training courses on pollution of coastal and ocean waters and their clean-up.
2. Supply of available equipment and technologies from developed countries to measure and control pollution.
3. Sharing of experiences and knowledge through exchange of visits, data and web portal.

4 National and regional training workshops to bring together expertise and experience from developing and developed countries, fostering South-South, North-South and triangular cooperation to access best available information and skills.
5 An integrated, well-planned capacity-building programme, to be backed by adequate financial and institutional support.

Based on the above introduction, let us examine two cases: the Baltic Sea, a non-UNEP-initiated programme; and the Mediterranean Sea programme, administered by the UNEP, and review what kind of capacity-building activities have been initiated in these two RSPs.

The Baltic Sea programme

The Baltic Sea is protected by the 'Convention on the Protection of the Marine Environment of the Baltic Sea Area' (1992), usually known as the Helsinki Convention. In 1974, for the first time ever, all the sources of pollution around an entire sea were made subject to a single convention, signed by the then seven Baltic coastal states. In light of political changes in 1990, and developments in international environmental and maritime law, a new Convention was signed in 1992 by all the nine states bordering on the Baltic Sea, and the European Community. The Convention covers the whole of the Baltic Sea area, including inland waters as well as the water of the sea itself and the sea-bed. Measures are also taken in the whole catchment area of the Baltic to reduce land-based pollution.

The governing body of the Convention is the Helsinki Commission – the Baltic Marine Environment Protection Commission – known as HELCOM. The present Contracting Parties to HELCOM are Denmark, Estonia, Finland, Germany, Latvia, Lithuania, Poland, Russia and Sweden. The European Union is also a party to this convention. In 2007, the Baltic Sea Action Plan was adopted, with four areas of priority:

1 Eutrophication – aiming at a Baltic Sea unaffected by eutrophication.
2 Hazardous substances –aiming at a Baltic Sea with life undisturbed by hazardous substances.
3 Biodiversity – aiming at the favourable conservation status of Baltic Sea biodiversity.
4 Aiming at a Baltic Sea with maritime activities carried out in an environmental friendly way.

The EU Strategy for the Baltic Sea Region and HELCOM

The EU Strategy for the Baltic Sea Region covers a wide range of issues, but foremost is the recovery of the Baltic Sea environment, led by HELCOM. Reports suggest that the Baltic Sea programme has proven successful in capacity-building efforts, directed mainly towards the five members known as the transition economies. (Vandeveer n.d.). The success factors are reported to be:

- adequate funding from multilateral and bilateral donors, as well as from the World Wildlife Fund, Coalition Clean Baltic, private sector funding and co-financing by member states;
- enthusiasm and commitment of transition country regimes, such as Estonia, Latvia and Lithuania;
- HELCOM serves as an effective regional coordinator;
- regular training, seminars and research studies undertaken jointly by member states. The people targeted for training were those who would be involved in the implementation and enforcement of the regime provisions.

However, Russia did not fare too well in terms of capacity building and participation, as it was relatively uninvested in the health of the Baltic Sea as a non-member of the EU (ibid.).

The Mediterranean Sea Programme

The main regulatory instrument to address the key environmental pressures and risks affecting the Mediterranean coastal and marine environment is the Barcelona Convention for the Protection of the Marine Environment and the Coastal Region of the Mediterranean and its Protocols. The Convention entered into force in 2004, replacing the 1976 'Convention for the Protection of the Mediterranean Sea Against Pollution'. The Barcelona Convention's main objectives are 'to prevent, abate, combat and to the fullest extent possible eliminate pollution of the Mediterranean Sea Area' and 'to protect and enhance the marine environment in that area so as to contribute towards its sustainable development'.

The Mediterranean Sea programme includes 21 countries and the EU. These countries, which have adopted the Barcelona Convention and its Mediterranean Action Plan, are Albania, Algeria, Bosnia/Herzegovina, Croatia, Cyprus, Egypt, France, Greece, Israel, Italy, Lebanon, Libya, Malta, Monaco, Morocco, Montenegro, Slovenia, Spain, Syria, Tunisia, Turkey and the European Community.

Another large collective effort working to protect Mediterranean marine and coastal environments is the UNEP/MAP GEF Strategic Partnership for the Mediterranean Sea Large Marine Ecosystem, or MedPartnership. The MedPartnership is led by UNEP/MAP and the World Bank,is financially supported by the Global Environmental Facility (GEF) and other donors, including the EU and other participating countries, and also encompasses other leading organizations (regional, international, and non-governmental) concerned with the Mediterranean's health. The MedPartnership works through two lines of action: technical and policy support led by UNEP/MAP (Regional Project); and project financing led by the World Bank (Investment Fund/Sustainable MED).

The Strategic Action Plan for the Conservation of Biological Diversity (SAP/BIO) was adopted by Contracting Parties to the Barcelona Convention in 2003. SAP/BIO (Strategic Action Plan for the Conservation of Biological Diversity) proposes a list of specific priority actions for Contracting Parties to undertake, such as inventorying, mapping and monitoring Mediterranean coastal and marine biodiversity; conserving sensitive sites, species and habitats; assessing and mitigating the impact of threats to biodiversity; developing research to improve knowledge regarding biodiversity; and developing skills to provide technical assistance, strengthen information sharing, encourage stakeholder participation and increase awareness.

Under the objective of developing research (SAP/BIO Objective 4/Target 3) to fill knowledge gaps on biodiversity, two priority actions (priority measures) were identified by the SAP/BIO, namely (1) improve and coordinate research on biodiversity; and (2) improve taxonomic expertise in the region.

In relation to capacity building, coordination and technical support, the Strategic Action Plan underlines two priority actions, namely, (1) establish a 'clearing-house' mechanism for on marine and coastal conservation activities; and (2) coordinate and develop common tools to implement National Adaptation Plans. Other work under the Strategic Action Plan includes expanding access to information for managers and decision-makers, promoting public participation, conservation of traditional knowledge and cultural values, facilitating international collaboration to raise region-wide public awareness, and enhancing environmental education for all, including journalists (UNEP 2015).

However, cooperation to preserve biodiversity in the Mediterranean has not proven very successful, largely because political, economic and cultural differences among European and Middle Eastern countries involved has hindered capacity-building efforts in the region. Vandeveer (n.d.) argues that efforts to develop capacity must focus not just on technical capacity, but also on institutional, administrative and political capacities. Initial diversity among countries' political and value systems therefore presents an obstacle to cooperative capacity-building initiatives.

In addition, reports suggest that efforts to conserve the Mediterranean's marine and coastal environments have faced a serious financial deficit in recent years (UNEP/MAP 2012). The contribution of the regional Trust Fund already had dropped around 20 per cent as of 2012 (Rochette & Billé 2012) and an extended functional review of the options to achieve financial sustainability was discussed during the most recent Conference of the Parties to the Barcelona Convention, held in December 2013.

Human rights law and capacity building

Initial developments

In historical perspective, the realization of political, civil, economic, social and cultural rights has proceeded sequentially, in stages, as evidenced by the adoption of global conventions on different aspects of human rights (Khan 2014). The first generation of rights concerned mainly political rights. The second generation of rights related to economic, social and cultural rights. At present, there is a strong movement for the realization of a third generation of human rights, which includes the right to a safe environment. This trend of expanding acknowledgement of a diverse set of human rights has made these rights an integral part of global and national development policies. From the 1993 Vienna World Conference on Human Rights to the latest world summit meetings on global development, including those held in Rio in 2012 and in New York in 2015, there has persisted a clear recognition that development and human rights are interdependent and mutually reinforcing.

This agenda of integrating human rights with development continues to be led by the UN system. In fact, the UN has been actively engaged in the process of mainstreaming human rights since 1997 and, in 2003, agreed on an interagency Common Understanding of a Human Rights-Based Approach to Development Programming. This Common Understanding highlights: the relationship between development cooperation, the Universal Declaration of Human Rights and international human rights instruments; the relevance of human rights standards for development programming; and the potential contribution of development cooperation to building the capacities of 'duty-bearers' and 'rights-holders' to realize and claim rights (OECD-World Bank 2003).

In 2007, following a 2006 study on integrating human rights into development, the OECD Development Assistance Committee published an Action-Oriented Policy Paper on Human Rights elaborating ten principles to help guide donors to effectively engage in human rights (OECD-DAC 2006; DAC 2007). These principles include, among others, the identification of areas of support to partner governments on human rights, backing for the demand side of human rights, and ensuring that the scaling-up of aid is conducive to human rights (OECD 2007). This report made evident that promotion of human rights in aid recipient countries has been made at least a soft conditionality by donors.

From the UN and donor perspectives, human rights work has both intrinsic and instrumental values (ibid.): it may be seen as an objective in its own right and as a means to improve the quality and effectiveness of development assistance. The intrinsic reasons for integration of human rights into development relate to moral and ethical imperatives and draw on the legal obligations emanating from the international human rights framework. This gives donors a norm-based organizing framework for development cooperation. Donors also focus on human rights for instrumental reasons, as a means to an end, to improve development and aid effectiveness related to governance and poverty reduction. Human rights frameworks help promote accountability of duty-bearers for their actions, as they empower citizens and marginalized communities to demand that the state respect, protect, and fulfil their rights. However, this approach contains potential for political conflict, as some partner countries may be less receptive to development cooperation linked to human rights. There are also practical challenges involved in operationalizing human rights in development programming.

Donor approaches to integrating human rights into development

In view of these considerations, donor approaches to integrating human rights into development programming can be sorted into five categories (OECD 2006b).

1 *Human rights-based approaches*: Human rights are considered as constitutive of the goal of development, leading to a new approach to aid and requiring institutional change and values.
2 *Human rights mainstreaming*: Initiatives are undertaken to ensure that human rights are integrated into all sectors of existing aid interventions.
3 *Human rights dialogue*: Promotion of foreign policy and aid dialogues include human rights issues, as a soft conditionality, where aid modalities and volumes may be affected in cases of significant human rights violations.
4 *Project approach to human rights*: Projects and programmes are directly targeted at the realization of specific rights, e.g., freedom of expression and group rights, or in support of human rights organizations to promote a civil society voice.
5 *Implicit human rights work*: Donor agencies may not explicitly work on a human rights agenda, preferring instead to use implicit descriptors like 'empowerment' or 'good governance'. However, the goals and content of their approach may still relate to other explicit forms of human rights integration.

Typically, multilateral agencies opt for less explicit approaches to integrating human rights into the development agenda of partner countries. But many bilateral agencies have explicit human rights policies as a political conditionality for aid delivery (Piron & De Renzio 2005). Since the early 1990s, the EU has introduced human rights clauses into its agreements, considering human rights, democracy, and the rule of law as 'essential elements' of development cooperation. As of December 2011, its human rights clause had been included in agreements with over 120 countries

(EC 2011). No reports on whether there has been cancellation of aid in cases of non-compliance with these clauses are available.

Tools to promote human rights

Based on a literature review, it is evident that capacity building targeted at the human rights sector has taken place both at national and international levels. At the national level, donors have traditionally supported human rights through projects that aim to build the capacity of human rights organizations, provide human rights training, or support the ratification of treaties and legal reform.

Some human rights projects are strongly research-based, providing analytical inputs and perspectives on a particular human rights issue. Discrete human rights projects often include research into a wide range of thematic issues linking human rights and development; surveys of needs and capacity gaps within countries, institutions, and sectors; analyses of indicators for measuring development outcomes using a human rights-based approach; and training materials on applying human rights-based approaches.

Capacity-building projects are often aimed at strengthening the ability of national institutions to promote the rule of law and human rights, such as support to ministries of human rights (in Burkina Faso), national human rights commissions (Uganda and Bangladesh), and ministries of justice (Mozambique).

There are a host of other tools that may be effectively used to promote human rights. Mexico, for example, has developed a web-based educational tool called Reforma-DH to educate its public about human rights. Similarly, in Paraguay, a web-based public policy tool on human rights information was installed. In Russia, a partnership of three universities has developed a Master's Programme in Human Rights. Bolivia has also started this kind of programme.. Morocco initiated a training programme on development indicators to improve measurement of compliance with international human rights. Cambodia has established a Fair Trial Rights Academy to train both lawyers and citizens on fair trial processes.

The multilateral development banks have developed implicit ways of safeguarding the basic rights of communities while designing development projects through adherence to environmental and social safeguard policies. The International Finance Corporation (IFC), for example, focuses on the role of private sector actors in fragile and weak states. With support from the Nordic Trust Fund, the International Committee of the Red Cross, extractive industry trade associations, and others, it launched an Implementation Guidance Tool for IFC's Voluntary Principles of Security and Human Rights in September 2011 (IFC 2011).

Multilateral agencies such as the UN system, the World Bank and others have historically not been well-equipped to carry out human rights-related capacity-building activities. In recent years, they have facilitated and participated in numerous learning events in order to better foster partnerships with academia, NGOs, the UN, EU, and the OECD relevant to human rights. These have included peer-to-peer exchanges with the OECD Development Assistance Committee's Human Rights Task Team (HRTT) or the UN, as well as training courses on human rights and development for fund grantees (Nordic Trust Fund 2010). The UN System Staff College (UNSSC) has also emerged as a provider of interagency training and learning within the UN system. It conducts a variety of training and learning activities in Turin, as well as at regional and country levels. Its work is organized around five areas, including human rights and development. The UNSSC, in close collaboration with the Office of the High Commissioner for Human Rights (OHCHR) and other UN agencies, offers its services to UN country team leaders and programme staff alike, with a view to building capacity to integrate human rights into all policy and programming processes.

The donors also fund research, networking and international events at the regional and global levels. For example, with the support of a Nordic Trust Fund (NTF) grant, financed by the governments of Denmark, Iceland, Norway, Finland and Sweden, the World Bank and the IFC recently published *Women, Business and the Law 2012: Removing Barriers to Economic Inclusion*, which examines how regulations and institutions in 141 economies distinguish between men and women in ways that affect capacity to work or to start and run a business (World Bank & IFC 2012). In collaboration with the Danish government, the Nordic Trust Fund also supported the publication of a World Bank study, *Human Rights Indicators in Development: An Introduction* (World Bank 2010a), which considers the significance of human rights indicators for development processes and outcomes by connecting standards and obligations with empirical data. Finally, in 2011, the Nordic Trust Fund supported the World Bank's international law study, *Human Rights and Climate Change: A Review of the International Legal Dimensions*, comprising a literature review of human rights and environmental issues (World Bank 2011).

Financial mechanisms dedicated to human rights

The International Covenant on Economic, Social and Cultural Rights lays out state obligations that are to be carried out 'individually and through international assistance and cooperation'. The Covenant thereby recognizes that many states will be unable to meet their obligations acting alone and will require international support. This provision has two dimensions: an obligation to receive and an obligation to provide.

In response, several funding mechanisms have been made available to provide direct support to capacity building for human rights in many different ways. The World Bank Nordic Trust Fund was established in 2008 with a five-year, $17 million fund financed by the governments of

Denmark, Iceland, Norway, Finland and Sweden. The NTF's objective is to help the World Bank develop an informed view on human rights. This internal learning programme supports activities that generate knowledge about how human rights relate to the Bank's analytical activities and operations.

The Voluntary Fund for Technical Cooperation (VFTC) in the Field of Human Rights was established in 1987 by the UN Secretary-General. It is funded by voluntary contributions and provides technical assistance to countries upon request. Projects are implemented within the framework of the Technical Cooperation Programme, administered by the Office of the High Commissioner for Human Rights. VFTC developed an online performance monitoring system, which is proving to be a critical contribution to UNHCHR's efforts to become a fully results-based organization. Now VFTC has a new Thematic Strategy – 'Widening the Democratic Space', under which its main focus is to build and strengthen national institutions and frameworks to implement human rights under international human rights law.

In 2015, the total expenditure of the VFTC amounted to $17.4 million. The Fund received a total of $13.1 million in voluntary contributions andthe VFTC's deficit at the end of 2015 was covered by other reserves. During this period, the Fund provided resources for technical cooperation to build strong human rights frameworks at the national level in 30 regions, states and territories.

However, the availability of funding for capacity building in a new area such as integrating human rights into development, remains quite poor overall. This has perpetuated weak capacity in many developing countries and LDCs to meet minimum human rights standards, as well as a widespread lack of awareness of human rights duties and claims. Conversely, many countries have worked to develop national human rights action plans without the support necessary to construct appropriate linkages to development plans and budgetary processes. Finally, even human rights-related capacity-building efforts that were financed have largely demonstrated an over-reliance on external consultants and UN volunteers, making lessons about rights-based programming unlikely to be internalized or sustained (OECD-World Bank 2003).

Disaster risk reduction, risk management and capacity building

Developments for capacity building

In recent years, natural disasters have increased in both severity and frequency (IPCC 2014). Recent estimates show that economic losses from disasters such as earthquakes, tsunamis, cyclones and flooding now reach an average of US$250–$300 billion a year; future losses are predicted to be even higher. However, annual disaster-related losses are not evenly distributed across geographies. Instead, losses in LDCs and SIDS are many times greater than those confronting the developed world. In fact, North America faces only 1.19 per cent of annual global disaster-related losses, while Europe and Central Asia experience less than 1 per cent (UNISDR 2015). In developing countries during the last decade, over 700,000 people have been killed, over 1.4 million injured, and about 23 million made homeless as a result of disasters (Sendai Framework for Disaster Risk Reduction 2015). The urgent need for capacity development to address disaster risk in vulnerable countries has been further underlined by recent major disasters, such as the Indian Ocean tsunami in December 2004, hurricanes Katrina, Rita and Nargis, and tragic earthquakes in China, India, Pakistan and Haiti.

Until the end of the last century, disaster management focused predominantly on emergency response, relief and rehabilitation. However, in recent years, this model has evolved towards a broader approach to disaster and risk governance, centring on *ex-ante* disaster risk reduction (DRR) and management. In other words, efforts emphasize prevention and preparedness rather than response and rehabilitation (IRGC 2005; Renn 2008; Paleo 2009; IRGP-IHDP 2010). Capacity development and disaster risk reduction are related by two core realities:

1 disasters can reverse hard-won development gains;
2 capacity or the lack thereof lies at the heart of reducing disaster risk.

In recognition of the linkage between capacity development and disaster risk management, the UN International Strategy for Disaster Reduction (UNISDR), launched in 2000 by the UN General Assembly, aimed at building disaster-resilient communities by promoting increased awareness of disaster reduction as an integral component of sustainable development, with the goal of reducing human, social, economic and environmental losses due to disasters. The Hyogo Framework for Action (HFA) (2005–2015) was then adopted by the global community. The HFA laid out three strategic goals:

1 Integration of disaster risk reduction into development policies and planning.
2 Development and strengthening of institutions, mechanisms and capacities at all levels, in particular at the community level, that can systematically contribute to building resilience to hazards.
3 Systematic incorporation of risk reduction approaches into the implementation of emergency preparedness, response and recovery programmes.

In addition, the five priorities set forth under the Hyogo Framework were as follows:

1 Ensure that disaster risk reduction is a national and local priority with strong institutional basis for implementation.
2 Identify, assess and monitor disaster risks and enhance early warning systems.
3 Use knowledge, innovation and education to build a culture of safety and resilience at all levels.
4 Mainstream disaster reduction in economic and social sectors.
5 Strengthen disaster preparedness for effective response.

These strategic goals and priority actions clearly indicate that capacity building for disaster risk reduction was given utmost priority under the Hyogo Framework for Action. In fact, the HFA contains at least 42 references to capacity building, capacity development, or needed capabilities.

Then, in 2016, at the Third UN Global Conference, the Sendai Framework for DRR (SFDRR 2015–2030) was adopted. The Sendai Framework introduced a number of innovations, the most significant being a strong emphasis on pre-emptive disaster *risk* management as opposed to disaster management. It also defined seven global targets for disaster risk reduction, including increased international cooperation through adequate and sustainable support, with an all-of-society, all-of-state approach and primary responsibility to prevent and reduce disaster risk assigned to states. The Sendai Framework set four priorities: (1) understanding disaster risk; (2) strengthening disaster risk governance to manage disaster risk; (3) investing in disaster risk reduction for resilience; and (4) enhancing disaster preparedness for effective response and in order to 'Build Back Better' in recovery, rehabilitation and reconstruction.

There was a clear recognition in the Sendai Framework that "Developing countries require an enhanced provision of means of implementation, including adequate, sustainable and timely resources, through international cooperation and global partnerships for development, and continued international support, so as to strengthen their efforts to reduce disaster risk." The Sendai Framework also stipulates that:

> In addressing economic disparity and disparity in technological innovation and research capacity among countries, it is crucial to enhance technology transfer, involving a process of enabling and facilitating flows of skill, knowledge, ideas, know-how and technology from developed to developing countries in the implementation of the present Framework.

As climate change is already increasing the frequency and severity of disasters, climate action, especially adaptation plans, should include strong disaster risk management components (IPCC 2012; SEI 2014; UNDP 2014). However, climate change is still considered to be weakly integrated with disaster risk management efforts in most countries (SEI 2014), with the exceptions of the SIDS in the Pacific (UNDESA 2014). Therefore, integrating the policy areas of disaster risk reduction and climate change adaptation remains both challenging and crucial (SEI 2014).

The Hyogo Framework and the Sendai Framework give a broad indication of the types of capacities still needed for disaster risk reduction across countries. A whole-of-society approach, i.e. developing capacities for disaster risk reduction as a society-wide endeavour that requires multi-stakeholder engagement and participation, is clearly warranted. Available literature on disaster risk reduction points to a number of experiences, tools and resources relating to capacity development that may prove especially useful moving forward:

1 *Developing policy and implementation frameworks*: According to the Hyogo Framework Monitor, over 100 countries now have dedicated national institutional arrangements for disaster risk management. As of 2014, more than 120 countries had undergone legal or policy reforms, over 190 had established focal points, and 85 had created national multi-stakeholder platforms since 2007 (UNDP 2014; UNISDR 2015). Institutional strengthening for disaster risk reduction at national and local levels has already been seen to reduce mortality, at least in those events for which early warning is possible: in Bangladesh, India and elsewhere, the number of lives lost from cyclones and tidal surges has fallen greatly in the past decade.
2 *Availability and use of data and sharing of information:* The web-based Hyogo Framework Monitor is regarded as a pioneering approach which empowers countries to record their disaster losses systematically at all levels. As a result, the 2015 Global Assessment Report (GAR15) was able to present systematic and comparable disaster loss data from 85 countries and territories, compared to 56 countries and territories in 2013, 22 in 2011 and only 13 in 2009. Notable progress has also been made in disseminating information through public awareness programmes. According to the Hyogo Framework Monitor, the number of countries with national disaster information systems and mechanisms for proactive information dissemination has increased markedly over the last two reporting cycles (2009–2013), with important regional differences. In Africa, lack of capacity and funding and poor internet connectivity are all cited as barriers, and many countries face issues of sustainability. In Asia, some low-income countries appear to have more advanced systems than high-income countries (UNISDR 2015).
3 *Spread of disaster risk management education and building of awareness*: Education, especially formal education, can prove a strong foundation for individuals' understanding of disaster risk. Better-educated communities suffer lower mortality and lower welfare impacts from disasters, particularly in terms of lost income (Garbero

& Muttarak 2013). Investments in education, and particularly in female education, have been shown to reduce vulnerability to disasters and should therefore be presented as a core strategic investment in disaster risk reduction (Muttarak & Lutz 2014). Over the last decade, a large number of educational materials, such as teachers' guides, relating to disaster risk have been produced in various languages and distributed globally. About 72 per cent of countries reporting through the Hyogo Framework Monitor indicate that disaster risk reduction is included in some way in the national educational curriculum, with rates of coverage in primary school curricula slightly higher than those for secondary and university or professional programmes. Only about a third of countries were able to report covering disaster risk at all educational levels, including professional education programmes (UNICEF & UNESCO 2014). However, the content and quality of educational materials on disaster risk reduction have not been rigorously reviewed, and the uptake by educational institutions is not monitored. As a result, it is difficult to measure what progress has been made and to determine the extent to which efforts to reform curricula with disaster risk reduction considerations have been successful (UNISDR 2015).

4. *Improved technical abilities for disaster response*: Specific areas of disaster response-related technical expertise, such as urban search and rescue, improved building codes, enhanced protection of health facilities, incorporation of risk-reduction approaches into recovery planning, and local level partnerships, have recently developed.
5. *Advisory services*: There is a wide network of multilateral institutions and forums as well as regional platforms that offer advisory services on disaster risk management to countries in need.
6. *Networking*: There is an array of international, regional and national civil society organizations and think-tanks devoted to advocacy and research on disaster risk management.

Financial support for DRM capacity building

International financial support for disaster risk management has thus far proven insufficient in amount and in focus on developing human and institutional capacities. Some countries have established domestic budget mechanisms to ensure that disaster risk reduction efforts have some level of guaranteed resources. Examples include percentage allocations by law in the Philippines or specially earmarked funds in Mexico (IFRC & UNDP 2014).

However, in most of the LDCs and SIDS, disaster risk management is dependent on resources from emergency and contingency funds, which are only replenished when large disasters occur. Several countries have reported insufficient resources to maintain even basic response capacities to the

Hyogo Framework Monitor. On average, only around 1 per cent of the total international development aid budget is allocated to disaster risk management (Kellett & Sparks 2012). Of this 1 per cent, about 70 per cent goes towards emergency response measures, which rarely prioritize reducing future risk. This dependence on emergency assistance generates a form of *humanitarian materialism* in which disasters themselves become commodities for consolidation of institutional resources and power (UNISDR 2015).

In recent years, a growing interest in risk financing has emerged, stemming from both disaster risk reduction activities and climate change-related efforts. International financial institutions and insurance, reinsurance and catastrophe modelling companies (Arnold 2008; World Bank 2008; Cummins & Mahul 2009; Muir-Wood 2011; GFDRR 2015) have begun extending risk financing mechanisms, such as insurance, contingency financing and catastrophe bonds, to nations and even in regional arrangements, such as the Caribbean Catastrophe Risk Insurance Facility.

The existing system of disaster risk reduction aid has both strengths and weaknesses (UNDP 2011b). Strengths include: substantial progress sensitizing countries to the need to initiate disaster risk management efforts and establishing national institutions for the purpose, good data availability on natural and climate disasters, and a strong presence of regional and global networks and partnerships, which allows swift mobilization of cross-sectoral and cross-country expertise.

There are also some basic weaknesses in the system:

- While it is widely recognized that local governments should play a critical role in disaster risk management, there is little evidence to suggest any concerted effort to strengthen their capacities outside of a handful of individual cases (UNISDR 2015).
- The prevailing practice of capacity 'substitution' places heavy reliance on external experts to undertake disaster risk reduction responsibilities and tasks. This leaves in-house and in-country capacities under-utilized and under-supplied and limits long-term impacts of interventions.
- Disaster risk reduction programming and project development can sometimes be overly ambitious, causing the basics of developing functional and technical capacities to be neglected.
- Interventions are often regarded as time-bound projects rather than as continuing programmes, raising issues of sustainability and ownership of the outcome of development efforts.

Capacity building under the Montreal Protocol

Developments in capacity building under the MP

In the mid-1970s, Dr Sherwood Rowling and his student Mario Molina detected that the ozone layer, which serves as a protective shield against the penetration of ultraviolet rays from the sun, was thinning out due to human use of ozone-depleting substances. Large ozone holes were detected in the stratospheric layer of the atmosphere over Antarctica and North America. Following further observation, scientific research, data collection and analysis, 197 countries ratified the Vienna Convention of 1985 to control and phase out ozone depleting substances (ODS). Then, in 1987, the Montreal Protocol to the Vienna Convention was signed, in response to additional scientific findings. Developing countries initially were not particularly eager to address the issue, because the problem of ozone shield rupture caused by use of ozone depleting substances affected mainly countries in the temperate zone, which are largely industrialized. Later, through the London Amendment, a Montreal Protocol fund, called the Multilateral Fund (MF) was created to help assist developing countries in the process of phasing out ozone depleting substances. Developing countries consuming less than 0.3 kg of ozone depleting substances per capita in a year were entitled to Multilateral Fund financial support, as well as to technical assistance. The MF support in recipient countries is implemented by the four implementing agencies (UNDP, UNIDO, UNEP and the World Bank).

From the very outset of Montreal Protocol implementation, institutional strengthening was emphasized and international assistance focused on strengthening national capacity to meet compliance targets. The focus on capacity building in developing countries was warranted because it enabled pursuit of the following ends:

- enhanced ability to provide a suitable climate in the country for the expeditious phase-out of ozone depleting substances (ODS);
- increased coordination, promotion and monitoring of the country's activities to phase out ODS;
- improved information systems for collection, analysis and dissemination relating to issues involved in protection of the ozone layer;
- enhanced facilitation of the exchange of information with other Parties and organs established by the Protocol;
- improved reporting systems for national data on ODS consumption;
- mobilization of industry through workshops, networking and sectoral associations;
- campaigns to raise public awareness through advertising and other promotional measures.

Accordingly, detailed steps were designed for capacity building and institutional strengthening through financial support and technical assistance to developing countries:

1 The first step began with preparation of a country programme, consisting of a commitment to take appropriate action to ensure compliance with the Protocol. A typical programme contained an analysis

of a country's current production and consumption of ozone depleting substances, together with a strategy statement and plan of action to be implemented by the government. The approval of the country programme by the Executive Committee (EC) of the Multilateral Fund was established as a precondition for financial assistance and institutional-strengthening projects.

2 To proceed with implementation, the next step was to establish a mechanism, usually called 'Ozone Cell/Unit', under the Ministry of Environment in a typical developing country. This cell, led by a senior government official, would facilitate expeditious implementation of projects for effective phase-out of ozone depleting substances and ensure effective liaison between the partner government and the Montreal Protocol Executive Committee, the Fund Secretariat, and the implementing agencies. This mechanism was supposed to continue until 2010, even if the phase-out was completed ahead of this date.

3 Finally, developing countries put institutional strengthening funds to use. Possible expenditures were specified as the following (UNEP 2017).

- office equipment to provide basic infrastructure for information processing and dissemination, as well as to improve ozone units' communication facilities;
- operational costs, for such purposes as post and telecommunications, stationery, maintenance of equipment and awareness-raising;
- personnel costs: incentive payments to ozone unit staff, paid at the rate of 100–200 per cent of the basic national salary level attached to the position.

Of the four involved agencies, UNEP was the primary operator of the technical assistance programme of the Montreal Protocol. It has helped more than 100 developing countries to prepare their national implementation strategies and management plans, assisted with institutional strengthening projects, and facilitated regional networking of national focal points. It has provided trainings targeting key stakeholders, ranging from policy-makers to customs officers to refrigeration technicians to farmers, in more than 80 countries. The result has been more empowered, better informed, and more effective focal points, known as National Ozone Units (NOUs). In response to the changing needs and priorities of developing countries over the course of the Montreal Protocol's compliance period (which began in 1999), UNEP has significantly evolved its programme strategy and delivery mechanisms. It has moved away from a project-by-project, activity-by-activity approach towards an integrated programmatic strategy, under which it provides direct policy advice for implementation. It has also regionalized the delivery of capacity-building programmes and services by placing its Regional Offices at the forefront of project implementation.

A number of innovative tools have been employed to build capacity under the Montreal Protocol. Examples include online trainings, initiatives emphasizing South-South cooperation, efforts leveraging the expertise of non-governmental organizations and institutions, and development of cross-cutting networks of centres of excellence. South-South cooperation has been used in more than 20 countries to help build capacity to comply with the Protocol, while a business-to-business (B2B) website, the first of its kind in the UN system, has recently helped minimize the production of new ozone depleting chemicals by encouraging the trade of existing ones (Raj n.d.).

Scientists have observed that the damaged stratospheric ozone shield is recovering as a result of the efforts of the world community to implement this treaty. Key factors behind the success of the phase-out of ozone depleting substances were undoubtedly the Montreal Protocol's emphases on awareness raising and capacity building.

Synthesizing regime experiences: commonalities and differences

This chapter has briefly reviewed the ongoing capacity-building efforts under five different global regimes. These efforts were undertaken through technical cooperation programmes of donor agencies. The regimes are: (1) trade capacity building (TCB); (2) capacity building under the Regional Seas Programme (RSP); (3) capacity building to integrate human rights into development; (4) capacity building for disaster risk management (DRM); and (5) capacity building for the phase-out of ozone-depleting substances under the Montreal Protocol.

The review shows that there are several commonalities and few differences among approaches and tools used in capacity building among these regimes. These are briefly explained as:

- Institutional development and strengthening are the focus of capacity building in all the regimes. Where this aspect was weak, the regime did not function effectively. The example is the Mediterranean RSP programme, where wide diversity of state regimes and value systems, different levels of development, and political instability in some states meant that public sector capacity could not reach a critical mass. On the other hand, the Baltic Sea Programme was successful because the compatibility of state regimes and value systems contributed to stronger, more effective capacity building. In like manner, institutional capacity building has been relatively successful under the Montreal Protocol, both at national and international levels, due to its detailed programme and implementation plans.

- Education, training and research aimed at human resource development and improving professional competence on a sustainable basis form another strong basis for building national capacity. This was evident across all the regimes, with a little less focus on education, but more focus on training and awareness raising in the ozone regime. The regimes on trade, human rights and disaster risk management particularly focused on formal education at different levels, including development of research and analytical capacities.
- Strong financial support behind the capacity-building efforts under the trade and ozone regimes contributed greatly to effective compliance with the regime provisions. The aid recipients of technical cooperation under these two regimes also proved relatively able to build their own capacities. Such adequate financial support was made available by the donor countries because of their direct interest in building capacity in developing countries. Promotion of trade capacity and the phasing-out of ozone depleting substances promoted both direct and indirect economic and health benefits not only in developing countries, but also in industrial countries. In the case of the human rights regime, this was also true to some extent because of its contribution to enhancing aid effectiveness. But the responses from developing countries were not the same across the board. Generally, financial constraints on capacity building in the regimes of human rights, disaster risk management and the Regional Seas Programme in the Mediterranean continue to inhibit effective capacity building.
- National ownership of capacity-building efforts is another key to its progress. In areas of mutual interest between donors and recipients, capacity-building efforts are relatively successful. For example, aid recipient countries with direct interests in the trade and ozone regimes owned more of capacity building efforts, which then proved more effective than under other regimes. In addition, where aid projects are viewed as donor-driven, which is often the case, country ownership usually does not grow.
- Networking, partnerships and sharing of experiences have contributed to capacity building success in all the regimes.
- Web-based tools have contributed substantially to capacity building under human rights and disaster risk management regimes. For example, Mexico has developed a web-based tool to sensitize citizens to human rights, which some other countries are replicating. The web-based HFA Monitor in the disaster risk management regime allows aggregation and consolidation of disaster-related data across the globe by allowing users to upload information on disasters and losses faced.
- Capacity substitution at national level by external experts and consultants works against the building of in-house and in-country capacity. This was especially evident in the case of the disaster risk reduction regime, where external experts played a relatively strong role. This

inhibited in-house and in-country capacity utilization and their further development. Many developing countries like Bangladesh have developed a fair amount of managerial and technical expertise in disaster risk management because of their age-old experiences of living with natural and climate disasters, and capacity building efforts should take advantage of this existing capacity.

Seven lessons to take away

1 The first lesson to take away is that sustainable capacity support at both the international and the national levels is the key. This support relates to both institutional arrangements and financing provisions.
2 Mutuality of interests in relevant areas helps cement the bond between donors and recipients. For this purpose, an expanded understanding of national interests among both groups of countries is warranted. This will facilitate ownership of efforts in recipient countries.
3 Political stability in developing countries is another important supporting factor in sustaining efforts aimed at capacity building.
4 User-friendly web-based tools can go a long way in capacity building, particularly when produced in the countries' own vernaculars.
5 Networking, partnerships and sharing of experiences have contributed to capacity building success across all the regimes.
6 Capacity-building initiatives are better owned when programmes are designed on a real participatory basis, and implemented with transparency.
7 Finally, education, training and awareness at every level aimed at human resource development are key to sustaining long-term capacity building. Here both formal and informal initiatives are warranted. In addressing climate change as a long-term issue, human resource development should be the central focus at both national and global levels. For that purpose, climate education must be mainstreamed within national curricula at every level of education.

References

Arnold, M. (2008). The role of risk transfer and insurance in disaster risk reduction and climate change adaptation. Policy Brief. Stockholm: Commission on Climate Change and Development. Available at: www.ccdcommission.org/Filer/pdf/pb_risk_transfer. pdf

Cummins, J. D., & Mahul, O. (2009). *Catastrophe risk financing in developing countries: Principles for public intervention*. Washington, DC: World Bank.

DAC. (2007). *DAC action oriented policy paper on human rights*. Paris: OECD.

Ehler, C. (2006). *A global strategic review. Regional Seas Programme, United Nations Environment Programme*. Available at: www.researchgate.net/publication/283225253_A_Global_Strategic_Review_Regional_Seas_Programme

Ehler, C., & Chua, T. E. (2006). The ecosystem approach to integrated ocean and coastal management. In B. Cicin-Sain (Ed.), *Reports from the Third Global Conference on Oceans, Coasts and Islands: Moving the global agenda forward*. Newark, DE: University of Delaware.

EIF. (2006). Task Force Report on the Enhanced Integrated Report 2006. Available at: www.enhancedif.org/en/document/task-force-report-eif-2006

European Commission. (2011). Human rights and democracy at the heart of EU external action – towards a more effective approach. (Joint Communication) COM(2011) 886 final. Available at: http://eeas.europa.edu/sites/eeas/files/2011_human-rights-annual

European Commission. (n.d.). Baltic Sea Programme. Available at: http://ec.europa.eu/environment/marine/international-cooperation/regional-sea-conventions/helcom/index_en.htm

Garbero, A., & Muttarak, R. (2013). Impacts of the 2010 droughts and floods on community welfare in rural Thailand: Differential effects of village educational attainment. *Ecology and Society*, *18*(4).

GFDRR. (2015). Investing in resilience. Available at: www.gfdrr.org/sites/gfdrr/files/publication/Investing-in-Resilience_1.pdf

GESAMP (Group of Experts on the Scientific Aspects of Marine Pollution). (1991). *The state of the marine environment*. London: Blackwell Scientific.

Hass, P. M., Keohane, R. O., & Levy, M. A. (eds). (1993). *Institutions for the Earth*. Cambridge, MA: MIT Press.

IFC. (2011). Voluntary principles on security & human rights. Available at: www.voluntaryprinciples.org/files/Implementation_Guidance_Tools.pdf

IFRC (International Federation of Red Cross and Red Crescent Societies) and UNDP. (2014). *Effective law and regulation for disaster risk reduction: A multi-country report*. New York: IFRC.

IPCC. (2012). *Managing the risks of extreme events and disasters to advance climate change adaptation: A special report of Working Groups I and II of the Intergovernmental Panel on Climate Change*. Cambridge: Cambridge University Press.

IPCC. (2014). *Climate Change 2014: Impacts, Adaptation, and Vulnerability: Report of Working Group II*. Cambridge: Cambridge University Press.

IRGC. (2005). Risk governance: Towards an integrative approach, White Paper No. 1. Geneva: International Risk Governance Council.

IRGP-IHDP. 2010. Integrated Risk Governance Project Science Plan. Integrated Risk Governance Project and International Human Dimension Program for Global Environmental Changes. Bonn, Germany.

JITAP. (2004). Progress report (1 November 2003 to 30 April 2004). Geneva, 17 May 2004.

Kellett, J., & Sparks, D. (2012). Disaster risk reduction: Spending where it should count. *Global Humanitarian Assistance*. Briefing Paper. Available at: www.globalhumanitarianassistance.org/wp-content/uploads/2012/03/GHA-Disaster-Risk-Report.pdf

Khan, M. R. (2014). *Toward a binding climate change adaptation regime: A proposed framework*. London: Routledge.

Lecomte, S. B. H. (2001). *Building capacity to trade: A road map for development partners insights from Africa and the Caribbean, ODI*. Available at: http://ecdpm.org/wp-content/uploads/2013/11/DP-33-Building-Capacity-Trade-Insights-from-Africa-Caribbean-2001.pdf

Muir-Wood, R. (2011). Designing optimal risk mitigation and risk transfer mechanisms to improve the management of earthquake risk in Chile. *OECD Working Papers on Finance, Insurance and Private Pensions*, *12*, 1.

Muttarak, R., & Lutz, W. (2014). Is education a key to reducing vulnerability to natural disasters and hence unavoidable climate change?. *Ecology and Society*, *19*(1), 42.

OECD. (2006a). *Trade-related assistance: What do recent evaluations tell us?* Paris: OECD.

OECD. (2006b). *Integrating human rights into development: Donor approaches, experiences and challenges*. Paris: OECD.

OECD (2007). *Human rights and aid effectiveness*, Paris. Available at: www.oecd.org/development/governance-development/38713028.pdf

OECD-DAC. (2006). *DAC guidelines and reference series applying strategic environmental assessment: Good practice guidance for development co-operation*. Paris: OECD.

OECD-World Bank. (2003). *Integrating human rights into development*. Washington, DC: World Bank.

OHCHR. (2015). *OHCHR annual report 2015*, Geneva. Available at: www2.ohchr.org/english/OHCHRreport2015/allegati/Downloads/1_The_whole_Report_2015.pdf

Paleo, U. F. (ed.). (2009). *Building safer communities: Risk governance, spatial planning and responses to natural hazards*. Vol. 58. Amsterdam: Ios Press.

Piron, L. H., & de Renzio, P. (2005). *Political conditionality in Africa: An empirical study into its design, use and impact: DFID 1999–2004: Report for DFID*. London: ODI.

Raj, S. (n.d.). Capacity building. Available at: http://cwm.unitar.org/publications/publications/cbl/synergy/pdf/cat1/statements/unep_dtie.pdf (accessed 14 December 2016).

Renn, O. (2008). *Risk governance: Coping with uncertainty in a complex world*. London: Earthscan.

Rochette, J., & Billé, R. (2012). ICZM protocols to regional seas conventions: what? why? how?. *Marine Policy*, *36*(5), 977–984.

SEI (2014): Can innovative learning tools help boost adaptive capacity in small islands? Available at: www.sei-international.org/-news-archive/2793-can-innovative-learning-tools-help-boost-adaptive-capacity-on-small-islands

Sendai Framework for Disaster Risk Reduction, UNISDR, Geneva. Available at: www.preventionweb.net/files/43291_sendaiframeworkfordrren.pdf

UNCTAD & UNDP (2006). *Trade capacity development for Africa*, New York & Geneva. Available at: http://unctad.org/en/Docs/ditctncd20066_en.pdf

UNDP (2014). Disaster Risk Governance During the HFA Implementation Period. Background Paper prepared for the 2015 Global Assessment Report on Disaster Risk Reduction. Geneva, UNISDR.

UNDP. (2011a). *Democratic governance thematic trust fund*. New York: UNDP.

UNDP. (2011b). *Strengthening capacities for disaster risk reduction: A primer*. Available at: www.preventionweb.net/files/globalplatform/entry_bg_paper~strengtheningcapacityfordrraprimerfullreport.pdf

UNDP (2014). Disaster Risk Governance During the HFA Implementation Period. Background Paper prepared for the 2015 Global Assessment Report on Disaster Risk Reduction. Geneva, UNISDR.

UNDP, & IFRC. (2014). Effective law and regulation for disaster risk reduction: A multi-country report summary. UN Development Programme and IFRC, June 2014. Available at: http://reliefweb.int/sites/reliefweb.int/files/resources/summary-report-final-single-page.pdf

UNEP. (2001). *Ecosystem-based management of fisheries: Opportunities and challenges for coordination between marine regional fishery bodies and Regional Seas Conventions*. UNEP Regional Seas Reports and Studies, No. 175, Nairobi: UNEP.

UNEP. (2016). *Regional oceans governance making, regional seas programmes, regional fishery bodies and large marine ecosystem mechanisms work better together*. Nairobi: UNEP. Available at: www.cbd.int/doc/meetings/mar/soiom-2016-01/other/soiom-2016-01-unep-06-en.pdf

UNEP. (2017). *Policies, procedures, guidelines and criteria (as at April 2017). Chapter X: Institutional strengthening. Multilateral Fund for the Implementation of the Montreal Protocol*. Nairobi: UNEP.

UNEP/MAP. (2012). Report of the 17th Ordinary Meeting of the Contracting Parties to the Convention for the Protection of the Marine Environment and the Coastal Region of the Mediterranean and its Protocols, Paris (France), 8–10 February 2012. UNEP(DEPI)/MED IG.20/8, 14 February 2012, §21.

UNEP/MAP. (2015). Initial gap analysis on existing measures under the Barcelona Convention relevant to achieving or maintaining good environmental status of the Mediterranean Sea, in line with the Ecosystem Approach. Available at: http://rac-spa.org/nfp12/documents/working/wg.408_05_eng.pdf (accessed 17 December 2016).

UN General Assembly. (2011). *Oceans and the Law of the Sea: Report of the Secretary-General*, 11 April 2011, §154. New York: UN.

UNICEF, & UNESCO. (2014). Non-formal education as a means to meet learning needs of out-of-school children and adolescents. Available at: http://allinschool.org/wp-content/uploads/2015/01/OOSC-2014-Non-formal-education-for-OOSC-final.pdf

UNIDO. (2002). Ozone friendly industrial development: 10 years of UNIDO in the Montreal Protocol, Vienna, Austria, V.02–55504. Available at: www.unido.org (accessed 8 October 2004).

UNIDO. (2009). *Manual on operations under multilateral environmental agreements, Vienna*. Available at: www.unido.org/fileadmin/user_media/Publications/Pub_free/Manual_on_operations_under_multilateral_environmental_agreements.pdf

UNISDR. (2015). *Making development sustainable: The future of disaster risk management. global assessment report on disaster risk reduction*. Geneva: UNISDR.

Vandeveer, S. D. (1999). Capacity building efforts and international environmental cooperation in the Baltic and Mediterranean regions. Conference proceedings: Saving the Seas: Developing Capacity and Fostering Environmental Cooperation in Europe. The Woodrow Wilson International Center for Scholars, Washington, DC, 14 May.

Vinod, T. (2006). Linking individual, organizational, and institutional capacity building to results. Capacity Development Briefs; No. 19. Washington, DC: World Bank.

World Bank. (1998). *Development, human rights and the World Bank*. Washington, DC: World Bank. Available at: http://siteresources.worldbank.org/

BRAZILINPOREXTN/Resources/3817166-1185895645304/4044168-1186409169154/08DHR.pdf
World Bank. (2004a). *Integrated framework for trade-related technical assistance addressing challenges of globalization: An independent evaluation of the World Bank's approach to global programs*. Washington, DC: World Bank.
World Bank. (2004b). *World Bank's approach to global programs: Phase 2 report, case study: the multilateral fund for the implementation of the Montreal Protocol*. Washington, DC: World Bank.
World Bank. (2008). Disaster risk reduction: good practice notes, July. Available at: http://siteresources.worldbank.org/CHINAEXTN/Resources/318949-1217387111415/Disaster_Risk_en.pdf = (accessed 12 December 2016).
World Bank. (2010a). *Human rights indicators in development: An introduction*. Washington, DC: World Bank. Available at: http://siteresources.worldbank.org/EXTLAWJUSTICE/Resources/HumanRightsWP10_Final.pdf
World Bank. (2010b). *Knowledge and learning for human rights and development: Nordic Trust Fund progress report*. Available at: http://siteresources.worldbank.org/PROJECTS/Resources/1171NTFReportProof8.pdf
World Bank. (2011). *Human rights and climate change*. Washington, DC: World Bank. Available at: http://siteresources.worldbank.org/INTLAWJUSTICE/Resources/HumanRightsAndClimateChange.pdf
World Bank-IFC. (2012). *Women, business and the law*. Washington, DC: World Bank. Available at: http://wbl.worldbank.org/~/media/WBG/WBL/Documents/Reports/2012/Women-Business-and-the-Law-2012.pdf
WTO. (1997). An integrated framework for trade-related technical assistance. Available at: www.wto.org/english/tratop_e/devel_e/framework.htm (accessed 15 December 2016).
WTO. (2001). Ministerial Declaration, 4th session, WT/MIN(01)/DEC/1. Available at: www.wto.org/english/thewto_e/minist_e/min01_e/mindecl_e.pdf (accessed 14 December 2016).
WTO. (2010). Enhanced integrated framework: greater trade capacity for poor countries, press release, 10 April. Available at: www.wto.org/english/news_e/pres10_e/pr601_e.htm (accessed 10 December 2016).
WTO. (2014). *Annual report 2014*. Available at: www.wto.org/english/res_e/booksp_e/anrep_e/anrep14_e.pdf
WTO. (2015). *Annual report 2015*. Available at: www.wto.org/english/res_e/booksp_e/anrep_e/anrep15_e.pdf
WTO. (2016). Financing of TRTA. Available at: www.wto.org/english/tratop_e/devel_e/teccop_e/financing_trta_e.htm (accessed 10 December 2016).

4 Needed

A capacity-building framework that's up to the task

Introduction: a new framework needed

We already live in a climate-changed world. Together with IPCC AR5 (2014), the latest report and study by NOAA (2016) and Hansen et al. (2016) corroborate this fact. However, recent politics and policy-making have not responded adequately to the latest science on climate change. The UNFCCC synthesis report on Intended Nationally Determined Contributions (INDCs) (2015) shows that even if the contributions to emissions reductions pledged in all the INDCs submitted by the Parties are realized, the world will witness close to a 3 °C temperature rise by the middle of this century. A 2016 study by a 12-member team, led by former NASA scientist James Hansen shows that global annual temperature is already 1.25 °C rise above preindustrial levels (1880-1920). So they argue for a 'massive extraction' of carbon dioxide from the atmosphere, an uncertain effort which will likely cost the world an astronomic sum of $104 trillion–$570 trillion. This is alarming news.

How do we respond to this formidable challenge? There exists an acute mismatch between the capacities of developed and developing countries to address the challenge. The LDCs and most of the SIDS are nano-emitters, yet suffer worst from climate change and have the least capacity to adapt, while the historically big emitters have the greatest adaptive capacity. Though capacity building as a primary goal of development cooperation is over a quarter of a century old, the results for climate action so far do not appear satisfactory (Kuhl 2009; Keijzer 2013). Technical cooperation as an approach to capacity building under development programmes and projects has come under serious criticism (Kanbur 2000). Technical assistance through externally funded development programmes has failed not just to develop sustainable capacity, but also has worked to erode existing capacity by pulling in external advisors and technical assistance personnel (Fukuyama 2004).

Most climate-related capacity-building work has been donor-driven, short-term exercises, initiated by a large number of bilateral and multilateral agencies, which leave no capacity-building 'system' behind. Further, there is no consensus yet established on what capacity building means or what it entails. As seen in Chapter 2 of this book, the UNFCCC in particular has repeatedly emphasized capacity building as an important piece of the intergovernmental process, but climate negotiators have never attempted to officially define it. Although a framework laying down the basic principles, priorities, guidelines and scope for capacity building was adopted back in 2001, in the years since, climate change has only intensified. In view of this fact, capacity building needs that match the complex nature of the problem must be reconsidered and carefully identified. Given that previous efforts have proven unsatisfactory, what should the basic parameters of capacity building be? What are the best possible tools and methodologies for capacity building? How should a framework that can create and sustain a capacity-building 'system' in the countries concerned be conceptualized? How should such a framework be implemented? Thus, this chapter is devoted to developing a capacity framework that is up to the task of addressing the challenge of mitigating climate change and adapting to its increasing impacts. To begin with, let us try to understand the *nature* of the problem, in order to conceptualize, first, the capacity needs.

Capacity needs that can address the peculiarities of the climate change problem

Climate change is the poster child of global diplomacy today. Since the end of the Cold War, perhaps no other issue has demanded so much time, energy and resources from the global community. It can rightly be regarded as an extremely wicked policy problem, which combines science, economics, politics, sociology and more. The roots of the problem's continued intractability can be traced both to conceptual and practical aspects of the issue. The practical aspects relate to its spatial and temporal dimensions: tropical regions, where the developing countries are located will be hit hardest and show a relatively weak capacity to adapt, while the rich historical emitters are located in temperate zones, with stronger adaptive capacity. As climate change is a stock, rather than a flow, problem, the impacts will manifest fully in the decades to come, and future generations are likely to suffer most. But the conventional neoliberal economic paradigm, which the Convention and the Paris Agreement uphold, does not encourage investment in solving a problem whose impacts will manifest far into the future.

Thus, the complex and wide-ranging nature of the climate problem make the issue of capacity extremely challenging. In the climate change policy arena, there are different categories of problems (Table 4.1), such as:

- known knowns – the impact of increased local temperature for longer periods on crop cycles;

Table 4.1 Known and unknown aspects of climate change

Categories of events	*Known consequence or probability*	*Unknown consequence or probability*
Known event	(I) Known knowns – increased local temperature for longer periods will affect crop cycles	(II) Known unknowns – rising ocean temperature may increase intensity of cyclones but frequency of occurrence not known
Unknown event	(III) Unknown knowns – indigenous person knows of a are pest that will thrive in a warmer climate but has not told the responsible authorities about it	(IV) Unknown unknowns: ex-post only – e.g. corroded sewer pipes in Melbourne due to reduced water flow in adaptive response to drought, nonlinear tipping points – Rumsfeld (2002) used this term (things we don't know we don't know) to explain military uncertainties

Source: Dobes, Gunasekera, Katzfey, & McInnes (2012).

- known unknowns – the rising ocean temperature may increase the intensity of cyclones but the frequency of occurrence is not known;
- unknown knowns – indigenous knowledge of a rare pest that will thrive in a warmer climate but responsible authorities do not know about it;
- unknown unknowns – corroded sewer pipes in a city due to reduced water flow in adaptive response to drought.

As discussed in Chapter 2, the seventh Conference of the Parties to the UNFCCC back in 2001 adopted the capacity-building framework for developing countries, with a separate work programme for the LDCs. The framework included principles, priorities and guidelines. It also has identified 15 priority areas for capacity needs in developing countries. Submissions by the Parties over time have also specified an array needs (UNFCCC 2001). This wide range of needs can be classified into three major categories (Khan et al. 2016):

- Better understanding of the nature of the climate problem as it pertains to any country – i.e. what might be the physical manifestations of a changed climate and its implications for various economic activities and livelihoods, such as impact on agriculture production, for ecological systems, such as forests, mangroves, coral reefs, and fisheries, and for human and societal well-being through weather-related disasters, heat stress, etc.?
- Ability to formulate and implement national actions to both help limit the scale of the problem through mitigation of greenhouse gas emissions, and to limit the impacts on human, ecological, economic and other societal sectors and systems through measures to mitigate risks and adapt to them.

- Analysis, building consensus on, and articulation of national interests in the UNFCCC climate negotiations and obligations as well as the broader array of international climate-related discussions and activities that now engage most countries (Sagar 2000).

The capacity needed to address these sets of needs begins with scientific research, including building better climate models and downscaling them so that they can yield locally relevant results, then analysing the potential interactions between the physical manifestations of climate change and the human, ecological, economic and infrastructural systems to understand how the real-world impacts may play out within countries in the coming years under different global mean temperature scenarios. This is important since climate impacts can differ greatly across regions as a result of significant variations in, for example, expected temperature rise, precipitation patterns and coastal storms. To illustrate, it is expected that the tropics will see a significant decrease in average precipitation while the higher latitude areas may see some increase; and while mean temperatures will rise overall, the increases will likely be greater in higher latitudes.

Furthermore, the resulting human, social, economic and ecosystem risks again depend on the nature of the local geophysical, biological and human systems and their vulnerability to climate changes. For instance, a coastal storm of the same magnitude may result in a very different outcome in Haiti compared to Florida and sea-level rise may have a more significant consequence for Bangladesh than the Netherlands. Thus, an assessment of these climate risks is an exercise in the judicious combination of natural sciences with the social sciences to unpack the various determinants of risk and their interactions with each other. As the Fifth Assessment Report (AR5) of the Intergovernmental Panel on Climate Change noted, '[c]limate change will amplify existing risks and create new risks for natural and human systems [which] are unevenly distributed and are generally greater for disadvantaged people and communities in countries at all levels of development' (IPCC 2014: 13).

Developing mitigation and adaptation responses to the climate challenge requires an additional set of capabilities. On the mitigation side, it requires an understanding of the key sources of current greenhouse gas emissions and how to mitigate the growth in these emissions in light of the developmental challenges and aspirations of the country. The latter is a particularly important and critical issue for developing countries, i.e. how to simultaneously manage the climate as well as other sustainable development challenges and economic growth imperatives. This will require a careful examination of possible feasible choices available to a country, which in turn depends on its geographical context and natural resources as well as human and institutional capabilities, on the technological frontier in various sectors, and the availability of other financial and technical resources that might be required.

Adaptation, on the other hand, offers its own set of challenges. Even if a very ambitious mitigation programme is undertaken today, antecedent deposition of GHGs will cause an impact for some time to come. So integrating climate-risk-mitigation strategies into development pathways (e.g. better coastal infrastructure planning), while very attractive, requires careful and forward-looking planning, which is not often a strength of developing countries. Additional approaches, such as putting in place disaster response and management systems, offer a complementary set of options to help reduce and better cope with the impacts of climatic events. Given such a complexity and diversity of options, defining the 'best way forward' is as much a consensus-building exercise as an analytical and strategic exercise, since it requires inputs from diverse stakeholders.

Translating mitigation and adaptation options into action necessitates an understanding of what implementation pathways might look like. Successfully implementing the preferred roadmap will then require assembling the appropriate multidisciplinary technical, financial, (public and private) organizational tools, and policy knowledge and resources.

For instance, a small, relatively poor country may choose to deploy solar photovoltaic (PV) systems to promote rural energy access, which may be a high priority for the country. In doing so, it may work with international partners to adapt imported technologies for local use, if there is not a large enough market to sustain local manufacturing, and to develop/implement business models that suit the local context and also create local livelihoods. In a country with a larger market and greater technological capabilities, the development of local manufacturing capacity – and even the possibility of participating in global value chains – may be a possibility. The successful undertaking of these functions requires a diverse set of capabilities and resources in various domains (technical, financial, organizational, policy), of various kinds (strategic, analytical, project planning and coordination, project implementation, monitoring, assessment and learning), and in a variety of actors (academia, think-tanks, firms and governments).

Engaging with the climate negotiations within the UNFCCC and the numerous related discussions in various other forums requires yet another set of capacities. These pertain to the ability to systematically draw upon knowledge from the other two domains mentioned above, to develop and articulate a national position that is consistent with the context, constraints and development objectives of the country, and to engage with the international arena for outcomes that lead to the achievement of the UNFCCC objectives, particularly the provision of adequate and appropriate international support to enable national action.

All this suggests that a formidable array of capacities is required in order for developing countries to be full and effective partners in the global effort to address climate change. While it goes without saying that local capacity is central to this process, it may be that not all the relevant capacity resides locally. For example, many LDCs likely will not be able to lead or manage a climate modelling effort aimed at understanding the manifestations of climate change within their countries. In this case, it will be crucial to ensure that international climate modelling capacity is adequately responding to the need to generate, for example, the downscaled scientific information that can be used as a starting point for climate risk assessment in these countries.

But even apprising what specific issues might require examination (e.g. changes in rainfall patterns) will require an understanding of local issues and priorities. Since climate risks result from the interactions of climate change with local physical, biological/ecological and human/societal systems, the need for local knowledge becomes critical. Here local (natural and social scientific) capacity will play a major role. Similarly, monitoring and observation of climate impacts may require both international and local capacity. But as we move towards issues such as prioritizing and implementing mitigation and adaptation options in the context of national development objectives, local capacity will play an increasingly central role, since an understanding of the local conditions takes primacy.

In other words, with upstream, relatively 'objective' processes relating to understanding of climate phenomena (such as climate modelling and other scientific research), the relevant capacity and processes may be delocalized, although informed by local context and needs. As we move towards developing an understanding of climate risks, local knowledge become more important, since broadly risk is the result of interactions between climate phenomena and local systems (whether physical, biological/ecological, or human). Here collaborations between industrialized and developing country actors can be quite fruitful.

But as we move further towards issues where subjective judgements become even more important – such as which development objectives to prioritize while choosing among mitigation options, for example, or what might be the most suitable way to implement an option – appropriate and adequate local capacity is critical, with external actors preferably playing only a supporting role (such as providing information on good practices of policies and business models elsewhere).

Notably, as capacity resides in humans and organizations, given the breadth and the complexity of almost any aspect of climate change, networks and institutions that enable and guide the flow of knowledge play an important role in both harnessing such knowledge and gathering multiple perspectives can become key.

To sum up, the key objectives of capacity building broadly are the development of appropriate human, institutional and systemic capacity to engage in these three activities (understanding the nature of the problem in different contexts and scales; understanding and navigating the 'solution' space, including international engagement; and implementing mitigation and adaptation solutions). A variety of actors, domestic and international, can contribute to this capacity development process through appropriate knowledge, skills, expertise and financial resources. As knowledge also undergoes a process of depreciation over time, with climate change knowledge being no exception, a constant renewal in light of findings from climate science will be needed.

Though capacity building as a concept and an activity is over a quarter century old, neither the development agencies nor the negotiators of the UNFCCC have tried to define it – what it actually means, or what it entails. However, in order to conceptualize a framework, we at least need to understand its broad parameters. The next section is devoted to understanding this concept.

What capacity building is

Capacity building or capacity development?

To understand what capacity building is, we need to understand why also the term 'capacity development' is used, and whether there are any differences between these terms or if they are used interchangeably. The World Bank is regarded as the initiator of the concept of capacity building, though later the development agencies started using the concept of 'capacity development'. Some view that there is no basic difference between these two terms (Vincent-Lancrin 2007), while others argue that there is: capacity building is regarded as starting from scratch, while capacity development is understood as having a base from which to start the process (Kuhl 2009; Pearson 2011a). Other literature also shows that there is a distinction between the two (JICA 2006; Lusthaus et al. 1999; UNDP 2009a). Freeman (2010: 17) argues that capacity building 'sits with the more traditional top-down approaches of knowledge transfer and technical cooperation. The word development, however, acknowledges existing capacities and the focus is on strengthening what is there already rather than starting something new.' In a similar vein, DAC (2006: 12) argues that:

> The phrase capacity development is used advisedly in preference to the traditional capacity building. The 'building' metaphor suggests a process starting with a plain surface and involving the step-by-step erection of a new structure, based on a preconceived design. Experience suggests that capacity is not successfully enhanced in this way.

Perhaps this is the reason why development agencies prefer the term 'capacity development'. It is true that capacity development indicates some pre-existing endogenous processes and addresses the need to support or facilitate processes that are already there. Chandy and Khasras (2011) argue that what links both these ideas together is the concept of 'building on existing capacities'. Thus, there is no universal agreement on which term should be used to explain the phenomenon. Since the UNFCCC uses 'capacity building', this book largely follows suit, though we will use both terms interchangeably.

Defining capacity building

By the early 1990s, capacity building had become a buzzword, a kind of organizing theme for development cooperation. Still, there is no universally accepted definition of capacity building (Enemark 2002). A consultation by the international Working Group on Capacity Development found that of the donor organizations surveyed, '60 percent ... did not have a common agency definition that was authorized or in common use throughout the organization' (cited in Enemark 2003). Furthermore, it found that 'Amongst the multilaterals interviewed, UNICEF, IFAD and the World Bank, there was no commonly accepted definition' (IWGCB 1998).

Different disciplines, such as institutional economics, organizational development or political economy define the term according to their own perspectives (Morgan 2006; Freeman 2010). Likewise, development agencies continue to define it in their own ways. Box 4.1 presents a wide array of definitions. Keijzer (2016) claims that some international consensus on capacity and its development is captured by the definitions (OECD 2006), shown in Box 4.1.

The definitions presented in Box 4.1 show more similarities than differences, the latter being more of semantics and focus. The UNDP definition appears more comprehensive than others, as it is goal-oriented, quite broad and brings in the element of sustainability, pointing to the time dimension that the OECD also openly brings in. The NORAD definition adds the element of stakeholder satisfaction in the process of capacity building, while the SDC and USAID definitions focus on improvement of performance and resource potential. The Rwandan President's very short definition of capacity as capacity to get things done is plain and to-the-point. All these interpretations point to the development of both individual and collective agencies. This approach aligns with Sen's concept of 'capabilities', the ability and freedom to do things and choose from alternatives (Sen 1997).

From the above definitions and discussion, some broad and common parameters can be gleaned about the characteristic features of capacity building:

- It is a dynamic, not static, process, adapting to evolving developments.
- It invokes change and improvement.
- It is not a one-off, quick-fix solution, but a long-term issue, requiring investment of time and resources.
- It involves people, organizations, institutions and society as a whole and their interactions, ultimately as society-wide development of capabilities.

Box 4.1 Definitions of capacity and capacity building/development

UNDP, 1997: 'The process by which individuals, groups, organizations, institutions and societies increase their abilities: to perform functions, solve problems and achieve objectives; to understand and deal with their development need in a broader context and in a sustainable manner.'
World Bank, 1998: '[C]apacity is the combination of people, institutions and practices that permits countries to reach their development goals ... capacity building is ... investment in human capital, institutions and practices.'
OECD 1996: 'Capacity is the ability of people, organizations and society as a whole to manage their affairs successfully.

- Capacity development is the process whereby people, organizations and society as a whole unlock, strengthen, create, adapt and maintain capacity over time.
- Support for capacity development refers to what outside partners – domestic or foreign – can do to support, facilitate or catalyse capacity development and related change processes.'

CIDA, 1996: 'Capacity building is a process by which individuals, groups, institutions, organizations and societies enhance their abilities to identify and meet development challenges in a sustainable manner.'
NORAD: 'Capacity development is a process by which individuals and organizations increase their abilities to successfully apply their skills and resources toward the accomplishment of their goals and the satisfaction of their stakeholders' expectations.'
SDC: 'The process to improve performance at the individual, organisational, network and broader system levels with the aim of increasing management and resource potentials.'
USAID: 'Approaches, strategies, or methodologies used by USAID and its stakeholders to change, transform, and improve performance at the individual, organizational, sector, or broader system level.'
President Paul Kagame of Rwanda, 2011: Capacity as 'the ability to get things done', which goes beyond formal qualifications and technical skills development to include the cultivation of intangible or 'soft' attributes such as the ability to drive change and to build processes, organizations, and institutions which can deliver public services over the long term.

- It is driven from the inside as an endogenous process and requires ownership.
- It strengthens existing capacities to ensure sustainability.

It is evident that the concept of capacity building remains resolutely impervious to strict definition. It is argued that this elasticity allows donors to fund otherwise difficult projects. But maintaining such a broad understanding gives rise to competing agendas and divergent interpretations of outcomes. Some country participants also use the term as a way of attracting additional resources and cooperating with International Development Association (IDA) preferences. The aim of such a perspective is to avoid defining capacity in ways that could limit its range and flexibility. So the operational utility of the concept comes from its ambiguity and lack of boundaries (Morgan 2006).

Contrary to the seemingly neutral nature of capacity development as conveyed by these definitions, capacity development creates winners and losers and is inherently political (Eade 2007; Ortiz 2013). We will discuss this issue later in this chapter.

Perspectives on capacity building

The perspectives on capacity building also differ, depending on the actors and agents. Capacity building as a concept is closely related to education, training and human resource development. Some practitioners and analysts continue to see capacity mainly as a human resource issue, as skill development and training at the individual level (Alsop & Kurey 2005). This 'capacity as training' perspective has a long-standing history and is still a widely held view both in development agencies and among governments (Morgan 2006). However, such a perspective of capacity building represents a narrow interpretation, such as staff development through formal education and training programmes to meet the deficit of qualified personnel in the short term (Enemark 2003).

This conventional understanding has changed over time towards a broader and more holistic view, covering both institutional and country-based initiatives, as was evident from the array of definitions in Box 4.1. Morgan (2006) argues that current perspectives range from the macro and the abstract, for example, the capacity of whole country to address the related issues, to the micro and the operational level, for example, the ability of personnel to communicate with internal and external stakeholders. Now there is a growing understanding that the scope of capacity issues goes beyond the usual training and technical assistance approach discussed below in detail. This perspective indicates an institutional and 'systemic' level, i.e. country-wide capacity building. This expanded perspective is being promoted particularly by the UN system, where capacity development has been elevated from a means of achieving something to the ultimate mission of development. The UN General Assembly Resolution (UN, A/RES/50/120 Art. 22) refers to the 'objective of capacity-building' as 'an essential part of the operational activities of the UN'.

This brings us to the question of whether capacity building is just a means to an end or an end in itself. Lusthaus et al. (1999) discuss this issue in detail. Some writers (Lusthaus et al. 1999, citing Fowler 1997) believe it to be important to build capacity as an end in itself, while others argue that capacity development should be worked on as a means towards sustainability (Lusthaus et al. 1999, citing UNDP 1996). At one level, it is referred to as the generally accepted 'central mission of development cooperation' (UN, E/1997/65: para 5), implying that capacity development is the goal in itself. In some other strands, the goal of capacity development remains at a more intermediate level – the capacity to achieve overall development of a country. Some definitions, including those of the Canadian International Development Agency (CIDA) and the United States Agency for International Development (USAID), consider that capacity development aims at enhancing the ability of individuals and institutions to meet development challenges (Box 4.1). In the USAID perspective, the final goal of capacity development is development itself. This understanding aligns with the UN documents and is linked to notions of 'sustainable development' (UN, E/1997/65: para. 12). Several UN documents specifically state that 'a vision of development and of the kind of society to be nurtured is a prerequisite (for CD)' (UN, E/1997/651Add.3: para. 8). Thus, the mission of capacity development depends on the country and agency perspectives. There is not yet any consensus on this issue. This book will promote both the perspectives.

As mentioned before, for the last decade and a half, capacity building has been kind of an organizing theme for continued development cooperation, both for the aid agencies, the governments in developing countries, as well as the NGOs, both in the Global North and the Global South (Morgan 2006). The concept is being used as an umbrella framework to package and legitimize a wide range of programmes and projects (Morgan 1998). 'If human development is the "what" of UNDP's mandate, then capacity development is essentially the "how"' (UNDP 2009b). Still the questions remain as to what kind of capacities to build to address the new challenges of climate change, how to build them, who will take the lead, what processes to follow for long-term sustainability, etc. All these issues are related to conceptualizing the ownership and partnership by developing country agents, in such joint activities, with development partners. Another key challenge is the knowledge ownership and management in promoting capacity building. Who owns the knowledge and skills relevant to climate change? How should knowledge and skills be managed? We will discuss these issues later in the chapter.

Types of capacity

The concept of 'capacity' normally evokes a sense of a comprehensive whole, an integrated understanding of what it entails. However, it would be useful to have a brief discussion of the types of capacities within an integrated whole, in order to gauge the likely scope and processes of their initiation in developing countries.

One analyst (Nanfosso 2011) broadly categorizes capacity building activities into three groups: (1) professional enhancement; (2) improvement of procedures; and (3) organizational strengthening. Others categorize types of capacity in different ways. Gagné et al. (1992) categorize five types of capacity: (1) verbal information (declarative knowledge); (2) intellectual skills (procedural knowledge); (3) cognitive strategies; (4) motor skills; and (5) attitudes. Pearson (2011a) elaborates the types of capacity, which appear appropriate too for addressing climate change. Different institutions and agencies differentiate types of capacity according to what fits their needs and profiles. As shown below, types of capacity are often distinguished under different headings such as 'hard and soft', 'technical/functional' and 'social/relational', 'tangible/intangible' or 'visible and invisible' (Table 4.2). The distinctions among them are often not clear-cut. There is now a growing recognition that soft capacities are the essential underpinning for other types of capacity to exist. These different types of capacity can be applied to individuals, groups, organizations and networks as well as to the systems and conditions in different enabling environments. Based on contexts, different capacities are needed for different reasons, in different combinations and measures. Most often capacities will be a mix of hard and soft components that fit the purpose and enable individuals, organizations, networks and broader social systems to carry out their development objectives. Support for the development of any capacity therefore needs to be approached in different ways. Based on the literature and authors' understanding, Table 4.2 represents three types of capacities: hard, semi-hard and soft.

Levels of capacity building

As is already clear, capacity building is a concept which is broader than individual education and training as well as institutional development since it includes an emphasis on the systemic environment and context within which individuals, organizations, institutions and societies operate and interact (Enemark 2003).

Most of the available literature (OECD DAC 2006a; CHF 2008; UNDP 2009; World Bank 2009a) suggests that capacity development occurs basically at three levels: individual, organizational and societal. However, some sources differentiate between organizational and institutional capacities (North 1990), while some other sources add capacity development at the global level (Pearson 2011a). In all these levels, the key to success is ownership of the process and products (Bolger 2000; Lafontaine 2000; Lopes & Theisohn 2003). The Commission for Africa (2005) emphasizes that developing countries must be the ones to lead the initiatives.

Table 4.2 Examples of hard, semi-hard and soft capacities

Hard	*Semi-hard*	*Soft*
Though some analysts argue that tangible resources like infrastructure, money, buildings, lab facilities, computers, equipment, etc. can be considered the material expression or product of capacity, but they are not capacity in and of themselves. But authors of this book argue that these hard capacities are vitally important for the functioning of semi-hard and soft capacities	Capacities that are generally considered to be technical, functional, tangible and visible • Technical skills, explicit knowledge and methodologies (which for individuals can be considered as competencies) • Organizational capacity to function: appropriate structures; systems and procedures for management, planning, finance, human resources, monitoring and evaluation, and project cycle management; the ability to mobilize resources • Laws, policies, systems and strategies (enabling conditions)	Capacities that are generally considered to be social, relational, intangible and invisible *Operational* capacities such as: • Organizational culture and values • Leadership, political relationships and functioning • Implicit knowledge and experience • Relational skills: negotiation, teamwork, conflict resolution, facilitation, etc. • Problem-solving skills • Intercultural communication *Adaptive* capacities such as: • Ability and willingness to self-reflect and learn from experience • Ability to analyse and adapt • Change readiness and change management • Confidence, empowerment and/or participation for legitimacy to act

Source: Pearson (2011a) and authors' adaptation.

Individual level

This level addresses the need for individuals as a collective to enhance their educational levels and skill sets to function efficiently and effectively within organizations and within the broader system. Human resource development (HRD) is the focus here, about assessing the capacity needs and addressing the gaps through adequate and appropriate education and training. Capacity assessment and development at this beginning level are considered the most critical, because individuals/citizens populate the organizations and systems. The dimension of capacity at the individual level includes the design of educational and training programmes and courses to meet the identified gaps within the skills base and with the number of qualified staff to operate the systems. Capacity at this level can be attained through formal and informal processes.

Organizational level

This level includes both formal organizations such as departments or agencies, private sector entities, NGOs or an informal organization, such as a civil society or community-based or volunteer organization. At this level, methodologies for capacity building examine all dimensions of capacity, hard, semi-hard and soft, including its interactions within the system, other entities, stakeholders and clients. The dimension of capacity at the organizational level should include areas, such as enabling infrastructures and logistics, mission and strategy, culture and competencies, processes, resources (human, financial, technical and information resources), etc.

Douglass North (1990) emphasizes that a distinction must be made between organizations and institutions. Although like institutions, organizations provide a structure to human interaction, conceptually, rules which are institutions must be differentiated from the players, i.e. organizations. In this sense institutions can be subsumed as representations in the systemic/societal level, because the policies, rules and norms make a system enabling or not.

Broader system/societal level

The highest level within which capacity initiatives may be undertaken is the system or enabling environment level. For development initiatives that are national in context, the system would cover the entire country or society and all subcomponents that are involved. For initiatives at a sectoral level, the system would include only those components that are relevant. The dimensions of capacity at systems level may include a number of areas such as policies, legal/regulatory framework, a management and accountability perspective, and the hard and soft resources available. In this sense, it is akin to institutional capacity, which refers to the means a country has at its disposal at the administrative levels, especially for the implementation of country-level policies, plans and strategies.

The general consensus among the development assistance organizations is that capacity development should have a comprehensive approach, as it refers to the individual, the organizational and the system levels (Nair 2003). The concept is based on the understanding that these three levels of society are inter-related, and any change is only possible if all three levels are taken into consideration. Thus, the HRD approach on the individual level, the institutional development approach at the organizational level and the new institutionalism approach at the system level have all been incorporated into the much more extensive capacity development concept (Morgan 2006).

Global level

Some analysts (Pearson 2011b) include the global level when discussing capacity building for global environmental problems including climate change, since this level includes interactions among all the levels for negotiations and finding solutions. Obviously, the international agencies like the UN system, the World Bank, other intergovernmental organizations, international NGOs, etc. now tend to serve as knowledge brokers (Chasek 2010) among other levels. But our discussions in Chapter 3 on capacity building under different regimes showed that they also sometimes need to build capacity before building capacities for others.

Dimensions of capacity

Capacity building as a sustainability process is looked at from three dimensions: capacity building, capacity utilization and capacity retention (Table 4.3). Experience in developing countries shows that together with appropriate capacity building, capacity utilization and capacity retention are

Table 4.3 Dimensions of capacity building and utilization at different levels

Levels	*Capacity building*	*Capacity utilization*	*Capacity retention*
Individual	Development of appropriate and adequate skills, knowledge, competencies and attitudes	Application of skills, knowledge, competencies in the workplace and where needed	Reduction of staff turnover through appropriate incentive structure, creation of conducive work atmosphere, facilitation of skills and knowledge transfer within institutions
Organizational	Establishment of efficient structures, processes and procedures	Integration of structures, processes and procedures in the regular workflows, valuing and respecting meritocracy	Regular adaptation of structures, processes and procedures
Systemic/ institutional	Establishment of enabling and adequate institutions, laws and regulations	Enforcement of laws and regulations for good governance; maintenance of institutional memory; drawing of relevant expertise on lien from other agencies	Regular adaptation of institutions, laws and regulations; devising of appropriate incentive structure, creation of conducive work atmosphere

Source: Adapted from Pearson (2011a).

equally important. Better quality expertise in different areas of climate change is not adequately available, and the skilled personnel have better opportunities to move around not just nationally but at the international level, with better incentives and packages. So it is very difficult to retain trained and skilled staff, given the low level of incentives, particularly in the LDCs. This problem is exacerbated by the underutilization of available expertise because of nepotism, political or clan considerations or because of the lack of transparency and the appropriate processes of head hunting and the lack of focus on meritocracy in many developing countries. These problems can be addressed to a large extent through development of an appropriate incentive structure, establishing a merit-based system and a constant and active process of HRD at higher secondary and advanced level through tertiary education. This requires adequate investment and constant inflow of other resources, particularly in the science and technical education sector.

Capacity-building principles

The UNFCCC Capacity Building Framework adopted under the Marrakech Accords at COP7 in 2001 contains the guiding principles of capacity building. Together, relevant literature has also been gleaned to find the set of principles to be followed in capacity building:

1. Capacity development as a locally driven process; no 'one-size-fits-all' approach – this means that capacity building has to fit each country's specific needs and circumstances.
2. Start from and build on existing capacities.
3. A continuous and iterative process, based on 'learning by doing'.
4. Efficient, cost-effective and programmatic approach.
5. Coordination of maximization of synergies in capacity building among the national and bilateral agencies and multilateral environmental agreements.
6. Application of traditional knowledge, skills and practices.
7. Shifting the balance from supply to demand-driven support to capacity building.
8. Participation of all stakeholders in a transparent manner.
9. Sustainability in capacity building at all levels.
10. Defining and measuring capacity results.

Approaches to capacity building

The term 'capacity building' evokes a connotation of something practical. Capacity as a concept is regarded as having a weak intellectual pedigree in the larger world of development. It comes with no accepted or tested body of theory (World Bank 2005).

However, in terms of approaches to building capacity, we can look at it both from theoretical and practical lenses. Nanfosso (2011) mentions two theoretical approaches to capacity building: development of capabilities along Sen's line of thinking (Sen 1997, 1999). Sen has established a close link between the theories of capacity and human capital, which proved to be useful in conceptualizing the process of capacity building. Sen's concept of capabilities can be viewed both from individual and group perspectives, as freedom to choose from alternative choices in the society. In order to acquire such capabilities, education and training serve as the foundation.

The human capital theory centring on education was developed by Becker and Schultz in the 1960s, which considers education as an investment (cited in Nanfosso 2011). Together, if education is appreciated for its own sake, as a consumption good, it becomes difficult to dissociate part of the human capital as investment for future returns from the part acquired for the sake of enjoyment. In either case, the quality of human resources is enhanced. We are more concerned here with education as an investment for capacity building, to do things as desired both at the individual and societal levels.

Chapter 2 elaborated on the evolution of the concept of capacity development since the 1950s and the 1960s when donors and academics focused on public sector institution building through human resource development (Kuhl 2009; OECD DAC 2006a). The traditional approach to building capacity had been the transfer of knowledge from the Global North to the Global South using technical cooperation (TC) as a funding mechanism of development agencies. However, TC was facing increasing critique due to poor results in many countries, with low returns (OECD DAC 2006a). Some have argued that it failed to enable developing countries to create their own sustainable capacities (Nair 2003). In the early 1990s, the UNDP and Berg (1993) led a review of TC approaches and found that, despite some successes, the sustainability of such efforts was lacking: As mentioned earlier, external agency-led support even caused harm to local capacity building because of its substitution effect.

Berg (1993) concluded that the supply-driven nature of TC led to poor local ownership and, therefore, lack of commitment. It was argued that there was a lack of understanding of the broader political and social contexts within which capacity-building initiatives had taken place (Morgan & Baser 1993; Bolger 2000). Instead of adapting to the country contexts and understanding the local culture, approaches were externally designed and implemented, which did not promote country commitment (Lusthaus et al. 1999; Bolger 2000; OECD DAC 2006a). So from the late 1990s a changed approach was called for. It apparently moved away from the conventional donor-led knowledge transfer approach to one of ownership through partnership and strengthening capacities (Lafontaine 2000; Kuhl 2009). Relationships were supposed to shift away from being donor-driven to a more collaborative partnership model where benefits were to be mutually shared (Horton et al. 2003). Instead of donors imposing their vision of development on poor countries, the focus became that of strengthening the capacity of local partners, including NGOs, who then can drive their own development. The World Bank (2009b) rightly argues that there is evidence to show that capacity is built faster when the process is endogenous.

These approaches were echoed in the 2005 Paris Declaration, the 2008 Accra Agenda for Action and the Busan high-level meeting on aid effectiveness. Since then, the new term of 'development effectiveness' has been introduced (Mawdsley et al. 2014). All these initiatives acknowledge that sustainability of development assistance is connected to local capacity, and that capacity building is 'an endogenous process of change' that strengthens these capacities. With ownership seen to be key to sustainable development, the focus is to support initiatives which are led by the recipient country. But the problem continues to persist, as is evident from the COP21 decision (Decision 1/CP21, para 73). The decision identified the problem of ownership as one of the nine elements of the five-year capacity building work programme of the PCCB. So COP21 asks to explore the ways and means of promoting country ownership of capacity-building efforts in developing countries.

Still, the problems continue to persist, as Keijzer (2016) emphasizes that development partners have made only half-hearted and timid steps to reform their approaches to supporting capacity development, and that a significant proportion of support fails to respond to the criticisms of the past, for example, by continuing to be supply-driven, tied, or insufficiently monitored and evaluated. This is the reason the indicators of the Paris Declaration monitoring framework relating to TC were judged least reliable (OECD 2006: 23–24). Initially, most criticism came from outside the development assistance organizations, but since the late 1980s it has increasingly also come from within (Berg 1993; Morgan & Baser 1993; UNDP 1993). The World Bank, which greatly influences development policy, itself questioned the practice of development assistance over the last four decades because the 'sustainable' results achieved are relatively meagre (World Bank 1998).

So the TC approach was disputed to such an extent that independent country programming aid analysis chose to exclude TC altogether (Kharas 2008). The resistance to reform and lack of effectiveness of capacity development support can also be explained by the underinvestment in HRD of many developing countries, and in particular in ensuring functional education systems. They are also linked to strong performance disincentives among civil servants and other administrators (Wohlgemuth 2005; Manning 2012). Therefore, some argue that the continued priority of capacity development in development cooperation 'is more a sign of previous failure than anything else' (Wohlgemuth 2005: 16). If countries had seriously invested in HRD, there would have been a better local market for technical assistance.

Literature suggests that when actors in developing countries accept development aid as a kind of service in the form of credits, subsidies, expert advice, training programmes, organizational development programmes, and so forth, they regard the relevant problems not as their own, but as issues to be solved by development aid (Jaycox 1993). So there is a growing sense that development should rather be understood as an endogenous process of transformation that must be upheld by the developing countries themselves (Kuhl 2009). The World Bank has therefore started advocating for project ownership based on a demand-driven approach whether the recipients truly want to have the programmes in the first place (Rottenburg 2002: 237). This is the context in which the concept of capacity development may be cast now (Hilderbrand 2002). This is the ever elusive issue to which we now turn our attention.

Ownership of capacity building through partnership

Origin of partnership

The critical review of traditional approaches to TC warrants a new and fresh approach for continuing development cooperation aimed at capacity building. As a result, a new organizing strategy was introduced at the beginning of this century, namely, promoting ownership through partnership as the *modus operandi* with recipient countries. Since then, partnership has become a ubiquitous term in development lexicon. The Accra Action Programme and the Busan high-level meeting on aid effectiveness specifically embraced the concept of partnership, while the goal of ownership of development was mentioned also in the Paris Declaration on Aid Effectiveness.

The jargon of 'partnership' in international development can be traced back to two international commissions: the 1969 Pearson Commission report on aid, entitled *Partners in Development*, and the 1980 Brandt Commission report *North-South: A Programme for Survival* (Maxwell & Riddell 1998: 258–259). Despite such efforts, foreign aid throughout the Cold War period remained dominated by geopolitical considerations Crawford (2003) elaborately traces this development. Later, a shift in inter-state relations was evident in the document of the OECD DAC, entitled *Shaping the 21st Century: The Contribution of Development Co-operation* (DAC 2006a). After a decade and a half of direct conditionality impositions in aid relations, it was emphasized that developing countries and their people must be responsible for their own development. Thus, the principle of 'locally-owned country development strategies based on an open and collaborative dialogue with civil society and external partners' was accepted (Crawford 2003).

This idea of 'partnership' has subsequently been embraced by a range of bilateral and multilateral agencies, such as the UK government's White

Paper on International Development (DFID 1997), and the World Bank's 'Comprehensive Development Framework' (World Bank 1999). The World Bank has enthusiastically adopted the discourse of 'partnership' as underpinning its new and integrated approach to development, which is articulated and 'owned' by the country concerned. Introduced by the World Bank President, James D. Wolfensohn, in his annual address to the Board of Governors in September 1998, the notion of 'partnership' was described as one 'led by governments and parliaments of the countries, influenced by the civil society of those countries, and joined by the domestic and international private sectors, and by bilateral and multilateral donors' (Wolfensohn 1998: 9).

The aid programmes to Africa by the European governments were dubbed in the language of 'partnership', for example *Partnership Africa* by the Swedish government (Swedish Ministry of Foreign Affairs 1997). The so-called new 'Marshall Plan' for Africa is entitled 'New Partnership for African Development' (NEPAD). Launched in October 2001 by African leaders, and endorsed by the African Union, NEPAD, however, continues to remain dependent on rich country support.

Why the change to a partnership approach?

Crawford (2003) presents two distinct explanations. The first points, as mentioned before, to a more pragmatic response by donor agencies to perceived shortcomings in aid performance, summarized as 'enabling more efficient use of scarce resources, increased sustainability and improved beneficiary participation' (Lister 2000: 228). The second explanation entails a more searching analysis of donor motivation and opens up two lines of enquiry. One relates to a defensive strategy by donor agencies to counter criticism of their activities, both at home and abroad. Fowler (2000: 1, citing Hudock 2000) refers to agency concerns over declining aid volumes in the post-Cold lWar era and the need to foster domestic support, while Lister (2000: 229, 235) points to a perceived need by donor agencies to legitimize their approach against calls for its reassessment by Southern critics. The other proposition points to a Machiavellian intent, articulated by Fowler: 'Partnership' may appear innocuous but can be 'a terminological Trojan Horse', as

> an instrument for deeper, wider and more effective penetration into a country's development choices and path ... By appearing to be benign, inclusive, open, all embracing and harmonious, partnership intrinsically precludes other interpretations of reality, options and choices without overtly doing so.

Thus, it serves to 'co-opt and sideline potentially opposing ideas and forces that express and propagate alternative views' (Fowler 2000: 7). Therefore, rather than relinquishing control to local actors, partnership can be a 'mystification of power asymmetry' (ibid.: 3) and 'a more subtle form of external power imposition' (ibid.: 7). Pender offers a similar interpretation of the World Bank's latest approach to development that partnership and the scope for national ownership are 'severely constrained' (2001: 409). He claims partnership as an instrument for 'seek[ing] to influence development in a far more all-encompassing way' (ibid.: 408). In his view, this amounts to a 'modified conditionality' under the World Bank's redefined concept of development. These discussions point to an imbalance in the power equation that favours the donor countries.

Partnership and power

Capacity building does not take place in a vacuum. It involves resources, both at home and on the international fronts, and hence it is related to the exercise of power by those who command such resources. In climate change-related capacity building, three crucial external resources are involved: money and knowledge and mitigation technology in particular. Obviously, how power over transfer of these resources is exercised between partners in decision-making is important for partnership analysis. Hyden (2008) argues that an understanding of power illuminates the challenges involved in transforming relations between donors and recipient governments as well as between governments and civil society organizations.

The concept of power is at the heart of political theorizing, yet it is one of many that remains contested. The importance of power relations in analysing institutional change was highlighted long ago, by Marx (1894), Gramsci (1971), Perroux (1973) and Galbraith (1976, 1984), for example. In the institutionalist tradition, Olson (2000) and North (1990) underline the role of interest groups in influencing and controlling the decision process. On the one hand, there is the intuitive understanding of power as domination, both overt and covert ways to manufacture consent to the choices of the hegemon (Ciplet et al. 2015).

When money, knowledge and technology transfer are involved as variables in any deal, achieving genuine partnership becomes extremely challenging, affecting mutually respectful partnerships. Some studies (Malhotra 1997; Gulrajani 2014; Mawdsley et al. 2014) show that decision process, transparency and accountability requirements have largely been a one-way track, rather than mutual. Any change from this fact remains bleak as long as accountability flows upstream, to donors, rather than flowing downstream, to development and the empowerment imperatives of the 'beneficiaries' of assistance and the broader communities of developing countries (Malhotra 1997).

In terms of knowledge as power, Stiglitz (1999: 308) cites Thomas Jefferson, the third President of the United States, who described knowledge as: 'he who receives an idea from me, receives instruction himself without lessening mine; as he who lights his taper at mine, receives light without darkening me.' This can be regarded as an early notion of knowledge as a public good. Stiglitz (ibid.: 308) argues:

> Today we recognize that knowledge is not only a public good but also a global public good. We have also come to recognize that knowledge is central to successful development. The international community, through institutions like the World Bank, has a collective responsibility for the creation and dissemination of one global public good – knowledge for development.

Knowledge as a public good is both non-rivalrous in consumption and non-excludable. This non-excludability of anyone from public goods causes their undersupply, because the private sector usually remains not interested in its provision, without adequate protection of their provision either by monopoly pricing or by patents. So, knowledge as an impure public good requires public support at both the national and global levels. Besides, global knowledge requires adaptation with local and indigenous knowledge, particularly to address the impact of climate change. Thus, though the industrial countries have the monopoly in R&D for the production of new knowledge, skills and technology, developing countries have contextual knowledge, based on their experiential learning. This kind of indigenous knowledge gained through life experiences and direct exposure to nature obviously can contribute to capacity building in adaptation also in the industrial countries. Therefore, we argue that knowledge to address climate change, the need for which arises from historical and current emissions of GHGs, should be regarded as a global public good (Khan, 2014), with global provisions for its funding including technology transfer.

In any case, let us look at the principles of partnership, which we can consider while analysing the implementation of capacity-building projects/programmes in developing countries and LDCs. This will allow us to gauge the extent of partnership, if any, for promoting country ownership as being established in the developing countries.

Principles of partnership

Since partnership and power are contested concepts, let us look at the principles of partnership, how it is exercised in real life while implementing capacity-building projects. Lister (2000: 228) identifies its core meaning as 'a working relationship that is characterized by a shared sense of purpose, mutual respect and the willingness to negotiate' (Pugh et al., cited in Buchanan 1994: 9). This definition emphasizes the values and principles espoused by those in a partnership endeavour. Common usage, however, leads to multiple interpretations. Comparison with the use of the term, 'participation' is instructive. There are two distinct conceptions of participation, described as 'instrumental' and 'genuine' (Brohman 1996: 252). Instrumental partnerships are used as means to achieve some ends by the development agencies, but retain control over activities and resources. In contrast, a genuine approach entails local communities setting the development agenda from the outset and remaining in control throughout. In this sense, participation is valued as an end in itself, whereas participation in the instrumental sense is more compatible with the status quo. In this vein, Maxwell and Riddell (1998: 260) refer to 'weak' and 'strong' models of donor-recipient partnerships. In the 'weak' version, partnership is limited to information sharing and policy dialogue, but in the stronger version, partnership is extended to joint decision-making in country programmes and financial agreements.

A review of the literature (Lister 2000; Crawford 2003) provides us with certain principles of partnership in aided projects: However, the principles of mutual respect, mutual accountability and transparency are not well explained. The principles of partnership are:

- shared purpose;
- mutual respect;
- mutual trust;
- mutual support;
- joint decision-making;
- mutual accountability;
- financial transparency;
- long-term commitment;
- respect for sovereignty and the right of national actors to determine their own policy options.

Financing for capacity building: a new approach

As education, training and public awareness serve as the foundational base of capacity building under Articles 11 and 12 of the Paris Agreement, their financing at all levels is extremely important. However, budget allocations for the education sector are obviously lower in the LDCs compared to other advanced developing countries. This phenomenon was shaped by external input. For example, one study by the World Bank on 98 countries covering the 1960–1997 period estimated that the social rate of returns was 18.9 per cent in primary education against a 10.8 per cent for higher education, so the World Bank advised that priority should no longer be given to higher education and, as a consequence, spending on higher education was reduced from 17 per cent in 1985–1989, to 7 per cent in 1995–1999 in African countries (Psacharopoulos & Patrinos 2002).

However, in 2005, the African Commission, set up by the British government, clearly suggested recognizing the value of higher education for development and increased its capacity in Africa by investing US$500 million per annum (and up to $3 billion over ten years) in scientific and technological centres of excellence. Later the World Bank (2011a) also recognized the need for a 'knowledge-based approach to development' and emphasized that more focus should be put on higher education. In any case, as mentioned, overall investment in the education sector is quite low.

It is here that a different kind of external funding for capacity building can leverage the needed domestic funding. Khan, a co-author of this book, has argued elsewhere (Khan 2014) that while GHG emissions are regarded as a negative externality, a global public bad (GPB), climate change impacts because of GHG emissions and undersupply of mitigation should be regarded as a GPB as well. Logically, together with mitigation, adaptation should also be considered as a global public good (GPG). In an age of global commons problems, the conventional conception of the public good (Samuelson 1954) should have an expanded interpretation. Funding for provision of such environmental global public goods should be externally sourced, beyond the ODA, by taxing the global public bads. This is the most fundamental principle of neoclassical economics, i.e. internalization of externalities to correct for the market distortions. As a logical follow-up, the nature of financing for capacity building to address climate change should be changed. It should no longer be regarded as a one-way charity, but a means of catering to a global good of capacity building to protect vulnerable countries and communities. Archibugi and Filippetti (2015) argue that the normative implication of the global public goods analysis in the case of knowledge requires greater public investment and international cooperation.

As mentioned in Chapter 2, there is no research yet on how much money is spent on capacity building for many different areas of development and environment, but loose estimates suggest that one-third to one-fourth of annual ODA goes to capacity building, and the overwhelming share is spent by bilateral agencies (Morgan 2006; Victor 2013). Since capacity building is a cross-cutting issue, often one component of larger projects, it is difficult to quantify total funding specifically dedicated to capacity building. In any case, funding for capacity building remains poor (Chen & He 2013; Nakhooda 2015; UNFCCC 2015). Wood et al. (2011) argue for reducing donor practices that undermine the development of sustainable capacity. This can be done by the promotion of untied budget aid (Kuhl 2009) on a long-term basis for education and capacity building to address climate change.

Further, with the emerging consensus that the traditional perspective on financing capacity building projects through technical assistance needs to be changed to a long-term funding cycle, successive declarations on aid effectiveness have agreed to a new approach to assistance, which emphasizes recipient ownership and greater control over the funds provided. As mentioned before, national public finance from industrial countries is not likely to allow ownership and control of projects by recipient countries through partnership, because money speaks loudly and thereby ensures donor domination and recipient country accountability upwards, to donors, rather than downwards, to vulnerable communities. Therefore, while funds flow from few international public sources, direct control of external resources by national governments is likely to remain limited. Ownership by recipient countries may be enhanced by extending the pool of available financing, perhaps, for example, by achieving agreement on global carbon pricing and creating a pool of levies. Whether the rich countries will live up to the commitments they agreed in Paris in 2005 is the million dollar question. Otherwise, the long-needed mutual trust will be lacking and vulnerable communities will continue to remain victims.

UNDP-ODI (2014: 26–30) proposes that in the post-2015 SDG agreement, national governments could commit to experimenting on a voluntary basis with coordinated innovative taxation schemes and use these taxes for international cooperation, because by not representing 'development aid' in the strict sense, aid 'will be less about "donors" and "recipients" and more about a common investment between equals in support of the common interest'. Capacity building to address climate change obviously represents such a common interest.

Universities: the central hub of capacity building

In our capacity building framework, we would establish universities in developing countries as the central hub of capacity building. Investing in universities is key to sustainable capacity-building systems under Articles 11 and 12 of the Paris Agreement (Hoffmeister et al. 2016). Historically, universities have proved to be powerful arbiters of knowledge in societies, with impact reaching well beyond their own boundaries (Winthrop & McGivney 2016). Universities have a ripple effect across all sections of society, running all the way from schoolchildren (through an influence on curriculum) to thought leaders and policy-makers. Finally, studies have shown that level of education is directly and positively correlated with economic growth and development (Nanfosso 2011; World Bank 2011b). For these reasons, universities in the developed world are uniquely well-positioned to lead and facilitate capacity building in the South.

While it is true that there are many knowledge generation sources outside of universities, including the private sector, civil society and NGOs,universities should be the hubs of capacity-building activities, the central pool of trained personnel and the main generator and disseminator of climate-relevant knowledge. Even the poorest countries have some universities. Political regimes come and go, but universities, like religious institutions, have proven extremely sustainable knowledge breeders across the world. Some universities in developing countries, for example, are already taking the lead in developing Master's programmes for students and professionals in climate change. Strengthening the existing programmes and helping develop new ones with the appropriate curriculum and research support would go a long way to building human resources to help tackle climate change. However, universities in developing countries, especially in the LDCs, largely lack resources, such as budgets to develop infrastructure, technical aids and internet facilities for learning, access to global knowledge and databases, library collections, research funds, etc. Overcoming these barriers requires funding and appropriate programme development to impart the specific skills needed to address climate change.

Capacity-building indicators

Resource scarcity and competitive claims on available resources are pushing the national and global development community to look for ways to ensure maximum bang for the buck spent on development interventions. In addition, increasingly widespread desire for enhanced accountability and transparency to achieve more effective development aid have led to calls to improve measurements of the performance of aided projects. Several global initiatives have been undertaken in recent years that focus on result-based management, including the goals and targets for the Millennium Development Goals (MDGs) and now the 17 newly agreed 17 Sustainable Development Goals, which span to the end of 2030. Besides, in a series of high-level meetings on aid effectiveness, donor agencies and partner countries have committed to becoming more results-oriented and 'managing for results' was adopted as one of the guiding principles of development cooperation (OECD 2003). Clearly, the development of metrics or indicators for measuring results has become a priority area in development effectiveness (UNDP 2009b; World Bank 2011b, 2014; Keijzer & Janus 2014).

There is as yet no agreed definition as to what a capacity-building indicator is. The OECD defines an indicator as 'a parameter, or a value derived from parameters, which points to/provides information about/describes the state of a phenomenon/environment/area with a significance extending beyond that directly associated with a parameter value' (1993: 6). This definition points out that indicators provide summary information of a phenomenon which can be communicated in simplified form to the different stakeholders of an intervention.

Many different types of indicators have been discussed in the literature. They mostly are indicators related to inputs, outputs, outcomes as short- and medium-term results and impact as long-term changes, both direct and indirect, in the area of intervention (Shyamsundar 2003; Holzapfel 2014). Efficiency, effectiveness and sustainability indicators may also be used to monitor and evaluate the efficacy of interventions (Holzapfel 2014).

The development of metrics for capacity building will help successfully realize Articles 11 and 12 of the Paris Agreement, as well as the global stocktake (under Article 14 of the Paris Agreement), and will support regular assessments of progress in capacity building. It also will serve as an analytical base for understanding patterns and determinants of successful capacity building. But measuring success in capacity building or its effectiveness is extremely challenging, as it involves both tangible and intangible or hard and soft results. Development agencies providing aid are often puzzled as to how to value results of interventions which often do not produce concrete measurable outcomes.

The sections discussed in this chapter lay out this puzzle. We have discussed different levels, types and dimensions of capacity building to address different needs, such as impact and vulnerability assessments, development of national plans and programmes, andh the articulation of national strategies for negotiations. As was evident from the past efforts discussed in Chapter 2, capacity building has not historically been effective and sustainable due to a lack of ownership of such interventions by the recipient countries. Productively, Article 11 of the Paris Agreement emphasizes this ownership. This framing chapter argued that a genuine partnership where the recipient countries can lead and hold some control over designing and implementation of capacity building programmes can ensure ownership in such efforts. The next logical question is how to measure such partnerships and ownership to ensure sustainability of such initiatives. As capacity building to address climate change is a new area, development of appropriate and robust metrics or indicators warrants rigorous applied research, with a learning-by-doing approach. We all agree that our goal is to develop a sustainable capacity-building system in developing countries, especially the LDCs, which will enable them to effectively address climate change. Keeping this and the framing issues in mind, let us present a sample of a few possible indicators of capacity building.

Sample of indicators related to imparting farmer adaptation training

- Input indicator measures material, human and financial resources used: budget allocated for the capacity-building programme of a group of farmers in a saline-prone area.
- Output indicators measure the goods and services that result from an intervention: number of farmers receiving adaptation training.
- Outcome indicators measure the short- and medium-term effects of the intervention's outputs: share of adaptation training graduates who were able to employ their newly-learned skills against climate change impacts.
- Impact indicators measure the long-term effects produced by the intervention, directly or indirectly: livelihoods of the impacted farmers in the saline-prone zones diversified and their income enhanced.

- Efficiency indicators represent the ratio of inputs needed per unit of output produced: amount of funding needed to impart adaptation training to a farmer.
- Effectiveness indicators show the ratio of outputs to produce one unit of outcome or impact: share of trained farmers who had been able to diversify their livelihoods and enhance their income.
- Sustainability indicators measure the persistence of outcomes or impacts over time after the intervention has ended: programmes of farmer adaptation training are institutionalized at local and national levels and continue after external funding and other supports end.

Synthesizing the framework

This framing chapter began by laying down the specific needs for capacity building that can effectively address climate change problems in the most vulnerable countries. Then we proceeded to explain what capacity building is, including its types, dimensions, perspectives, levels and principles. We found that the concept of capacity building has historically been constructed vaguely and capaciously, such that many possible meanings coexist under the single heading of 'capacity building.' Therefore, we have tried to lay out some basic parameters of capacity building.

As we have seen, capacity-building efforts to date have been haunted by the same old problems that continue to bedevil the process: it is short-term, project-based, consultancy-led, and donor-driven. These characteristics continue to stand in the way of ownership by the recipient countries. In recognition of these pitfalls, some reform has been initiated in the donor countries, but it remains half-hearted and cosmetic. On the other hand, lack of capacity in vulnerable countries is compounded by lack of commitment to a process funded externally. In order to overcome these problems, a new ideal of partnership between the funders and the recipients has been introduced over the last decade. The literature currently available on this approach does not speak positively of its effectiveness so far, but the approach has great potential.

The problem in partnership begins with external funding and knowledge transfer, either as software or hardware. Invariably these resources bring the exercise of power into play, often marring the partnership relationships. To attend to these vitiations, we have argued that funding for climate change science and policy education needs to be increased domestically, while external funding for capacity building must not be regarded as voluntary, but as a means of pursuing the global good of mitigation and protecting the many countries and populations who suffer worst from climate change despite being least responsible for causing it. This funding should take the form of untied budget support for education and capacity building on a long-term basis. In addition, knowledge and information needed for the purpose of combatting climate change are global public goods, so they should be shared by developed and developing countries. Vulnerable countries have rich indigenous knowledge, based on age-old experiential learning about adaptation, that should be shared for mutual advantage. Finally, sufficient international public finance should be mobilized and delivered through agreed mechanisms in order to contribute to recipient country ownership of capacity-building efforts and real partnership building.

In synthesizing our framework, let us conclude with Stiglitz's argument that successfully meeting the challenges posed by knowledge externalities depends critically on cooperative efforts at the international level. University partnerships across the world can spearhead these cooperative endeavours across borders.

References

Alsop, R., & Kurey, B. (2005). *Local organizations in decentralized development: Their functions and performance in India.* Washington, DC: World Bank.

Archibugi, D., & Filippetti, A. (2015). Knowledge as global public good. In D. Archibugi, & A. Filippetti (eds), *The handbook of global science, technology and innovation.* Chichester: John Wiley & Sons, Ltd.

Becker, G. (1975). *Human capital* (2nd edn.). Chicago: Chicago University Press.

Berg, E. (1993). *Rethinking technical cooperation* (4th edn.). Orlando, FL: Brace Jovanovich.

Bolger, J. (2000). Capacity development: why, what and how. In *Capacity Development*, Occasional Series 1(1). Gatineau: CIDA.

Brohman, J. (1996). *Popular development.* New York: Wiley.

Buchanan, J. (1994). Ethics and economic progress. *Journal des Economistes et des Etudes Humaines*, *5*(1), 1–9.

Chandy, L., & Khasras, H. (2011). Why can't we all just get along? The practical limits of international development cooperation. *Journal of International Development*, *23*, 739–751.

Chasek, P. S. (2010). *Confronting environmental treaty implementation challenges in the Pacific Islands.* Honolulu: East-West Center.

Chen, Z., & He, J. (2013). Foreign aid for climate change related capacity building. WIDER Working Paper. No. 2013/046, April.

CHF. (2008). Capacity development: Key to North-South NGO partnerships? CHF partners in rural development, February. Available at: www.chfpartners.ca/publications/documents/CapacityBuildingCaseStudyFinalFeb408.pdf (accessed 1 September 2009).

CIDA. (1996). *Capacity development: The concept and its implementation in the CIDA context.* Hull, Quebec: CIDA.

Ciplet, D., Roberts, T., & Khan, M. (2015). *Power in a warming world: The new politics of climate change and the remaking of environmental inequality* Cambridge, MA: MIT Press.

Commission for Africa. (2005). Our common interest: Report of the Commission for Africa. March 2005. Available at: http://allafrica.com/sustainable/resources/view/00010595.pdf (accessed 5 October 2009).

Crawford, G. (2003). Partnership or power? Deconstructing the Partnership for Governance Reform in Indonesia. *Third World Quarterly*, *24*(1), 139–159.

DAC. (2006a) *Shaping the 21st century: The contribution of development co-operation*. Available at: www.oecd.org/dac/2508761.pdf

DAC (2006b). *DAC in dates: The history of the Development Assistance Committee*, Paris: OECD.

DFID. (1997). *Eliminating world poverty: A challenge for the 21st century*. London: HMSO.

Dobes, L., Gunasekera, K, J., & McInnes, K. (2012). Adaptation to climate change: formulating policy under uncertainty. Sydney: Centre for Climate Economics and Policy (CCEP), ANU, Working Paper 1201. Available at: http://dev.cakex.org/sites/default/files/documents/CCEP1201Dobes.pdf

Eade, D. (2007). Capacity building: Who builds whose capacity?. *Development in Practice*, *17*(4–5), 630–639.

Enemark, S. (2002). Strengthening institutional capacity in land administration: Towards developing methodological guidelines. In Proceedings of FAO Workshop, The Land Tenure Service, Rome, 14–15 November.

Enemark, S. (2003). Understanding the concept of capacity building and the nature of land administration systems. FIG Working Week, April, Paris, France.

Fowler, A. (1997). *Striking a balance: A guide to enhancing the effectiveness of non-governmental organizations in international development*. London: Earthscan.

Fowler, M. (2000). Beyond partnership: Getting real about NGO relationships in the aid system. *IDS Bulletin*, *31*(3), 1–13.

Freeman, K. (2010). Capacity development theory and practice, Master's thesis, Oxford Brooks University.

Fukuyama, F. (2004). *State-building: governance and world order in the 21st century*. Ithaca, NY: Cornell University Press.

Gagné, R. M., Briggs, L., & Wager, W. (1992). *Principles of instructional design*. New York: Holt, Rinehart & Winston.

Galbraith, J. K. (1976). *The affluent society*. New York: New American Library.

Galbraith, J.K. (1984). *The anatomy of power*. New York: Hamish Hamilton.

Gramsci, A. (1971). *Selections from the prison notebooks* (ed., trans. Q. Hoare, & G. Nowell-Smith). London: Lawrence and Wishart.

Gulrajani, N. (2014). Organising for donor effectiveness: An analytical framework for improving aid effectiveness. *Development Policy Review*, *32*(1), 89–112.

Hansen, J., Sato, M., Hearty, P., Ruedy, R., Kelley, M., Masson-Delmotte, V., ... & Velicogna, I. (2016). Ice melt, sea level rise and superstorms: Evidence from paleoclimate data, climate modeling, and modern observations that 2 °C global warming could be dangerous. *Atmospheric Chemistry and Physics*, *16*(6), 3761–3812.

Hilderbrand, M. (2002). Capacity building for poverty reduction: Reflections on evaluations of UN system efforts. Unpublished MS thesis. Harvard University.

Hoffmeister, V., Averill, M., & Huq, S. (2016). *The role of universities in capacity building under the Paris Agreement*. Policy Brief. Dhaka: International Centre for Climate Change & Development (ICCCAD).

Holzapfel, S. (2014). Presenting results in development cooperation: Risks and limitations. Briefing Paper 4. Bonn: German Development Institute/DIE.

Horton, D., Alexaki, A., Bennett-Lartey, S., Brice, K. N., Campilan, D., Carden, F., ... Watts, J. (2003). Evaluating capacity development, experiences from research and development organizations around the world, Chapter 5. Canada: IRDC. [online] Available at: www.idrc.ca/en/ev-43625-201-1-DO_TOPIC.html (accessed 1 September 2009).

Hudock, A. C. (2000). NGOs' seat at the donor table: Enjoying the food or serving the dinner? *IDS Bulletin*, *31*(3), 14–18.

Hyden, G. (2008). After the Paris Declaration: Taking on the issue of power. *Development Policy Review*, 26(3), 259–274.

IPCC. (2014). *Fifth assessment report of the Intergovernmental Panel on Climate Change*. Edited by Core Writing Team, R. K. Pachauri, & L. A. Meyer. Oxford: Oxford University Press.

IWGCB. (1998). *Southern NGO capacity building: issues and priorities*. New Delhi: Society for Participatory Research in Asia.

Jaycox, E. V. (1993). Capacity building: The missing link in African development. Address to the African-American Institute Conference on Capacity Building, Reston, VA.

JICA. (2006). Summary: towards capacity development (CD) of developing countries based on their ownership. Concept of CD, its Definition and its Application in JICA Projects. September. JICA.

Kagame, P. (2011). Capacity building must focus on positive change, *The New Times*, 10 February. Available at: www.newtimes.co.rw/section/read/28328/

Kanbur, R. (2000). Aid, conditionality and debt in Africa. In F. Tarp (ed.), *Foreign aid and development: Lessons learnt and directions for the future* (pp. 409–422). London: Routledge.

Keijzer, N. (2013). *Unfinished agenda or overtaken by events? Applying aid and development effectiveness principles to capacity development support*. Bonn; Deutsches Institut für Entwicklungspolitik.

Keijzer, N. (2016). Open data on a closed shop? Assessing the potential of transparency initiatives with a focus on efforts to strengthen capacity development support. *Development Policy Review*, *34*(1), 83–100.

Keijzer, N., & Janus, H. (2014). *Linking results-based aid and capacity development support: Conceptual and practical challenges*. Bonn. DIE.

Khan, M. R. (2014). *Toward a binding climate change adaptation regime: A proposed framework*. London: Routledge.

Khan, M. R. (2016). Climate change, adaptation and international relations theory. *Environment, Climate Change and International Relations*, 14. Bristol: E-International Relations Publishing.

Khan, M., Sagar, A., Huq, S., & Thiam, P. (2016). *Capacity building under the Paris Agreement*. Oxford: European Capacity Building Initiative.

Kharas, H. (2008). The new reality of aid. In L. Brainard, & D. Chollet (eds), *Global development 2.0: Can philanthropists, the public and the poor make poverty history*? (pp. 53–73). Washington, DC: Brookings Institute.

Kuhl, S. (2009). Capacity development as the model for development aid organizations. *Development and Change*, *40*(3), 551–557.

Lafontaine, A. (2000). *Assessment of capacity development efforts of other development cooperation agencies*. New York: GEF-UNDP.

Lister, S. (2000). Power in partnership? An analysis of an NGO's relationships with its partners. *Journal of International Development*, *12*, 227–239.

Lopes, C., & Theisohn, T. (2003). *Ownership, leadership, and transformation: Can we do better for capacity development*? London: Earthscan.

Lusthaus, C., Adrien, M. H., & Perstinger, M. (1999). Capacity development: Definitions, issues and implications for planning, monitoring and evaluation. *Universalia Occasional Paper*, *35*, 1–21.
Malhotra, K. (1997). Something nothing words: Lessons in partnership from southern experience. In North South Institute, *Essays on Partnership in Development*. Ottawa: Renouf Publishing.
Manning, R. (2012). Aid as a second-best solution: Seven problems of effectiveness and how to tackle them WIDER Working Paper No. 2012/24).
Marx, K. (1894). *Capital: A critique of political economy*. Vol. III: *The process of capitalist production as a whole*. 1909 trans. of 1894 ed. by E. Untermann.
Mawdsley, E., Savage, L., & Kim, S. (2014). A 'post-aid world'? Paradigm shift in foreign aid and development cooperation at the 2011 Busan High Level Forum. *The Geographical Journal*, *180*(1), 27–38.
Maxwell, M., & Riddel, R. (1998). Conditionality or contract: Perspective on partnership for development, *Journal of International Development*, *10*, 257–268.
Morgan, G. (1998). *Creative organization theory*. Newbury Park, CA: SAGE.
Morgan, P. (1998). Capacity and capacity development. Some strategies. Note prepared. Hull, Quebec: CIDA.
Morgan, P. (2006). *The concept of capacity*. European Centre for Development Policy Management. Available at: http://preval.org/files/2209.pdf
Morgan, P., & Baser, H. (1993). *Making technical cooperation more effective: New approaches in international development*. Hull, Quebec: CIDA, Technical Cooperation Division.
Nair, G. (2003). Nurturing capacity in developing countries: From consensus to practice. Capacity Enhancement Briefs, Number 1 November: World Bank Institute [Online] Available at: http://info.worldbank.org/etools/docs/library/82361/CEbrief-01_Nov03%20-%20Nurturing%20Capacity%20in%20Developing%20Countries.pdf (accessed 14 September 2009).
Nakhooda, S. (2015). Capacity building activities in developing countries. Workshop on potential ways to enhance capacity building activities. Bonn, 17 October.
Nanfosso, T. R. (2011). The state of capacity building in Africa. *World Journal of Science, Technology and Sustainable Development*, *8*(2/3), 195–225.
NOAA. (2016). *Global analysis*. August. Available at: www.ncdc.noaa.gov/sotc/global/201608
North, D. C. (1990). *Institutions, institutional change, and economic performance*. Cambridge: Cambridge University Press.
OECD. (1993). *OECD core set of indicators for environmental performance*, Environment Monographs # 83, Paris: OECD.
OECD. (2003). *OECD framework for environmental indicators*. Paris: OECD.
OECD. (2012). Indicators on capacity building for adaptation, Nicolina Lamhauge OECD Environment Directorate. Available at: www.oecd.org/env
OECD. (2006a). *DAC guidelines and reference series applying strategic environmental assessment: Good practice guidance for development co-operation*. Paris: OECD.
OECD. (2006b). *Survey on monitoring the Paris Declaration of collective action*. Paris: OECD.
OECD. (2010). *Inventory of donor approaches to capacity development: what we are learning*. Paris: OECD. Available at: www.oecd.org/dataoecd/50/12/42699287.pdf

OECD. (n.d.). Brief: Paris Declaration on Aid Effectiveness (2005) and Accra Agenda for Action (2008). Available at: www.oecd.org/dataoecd/11/41/34428351.pdf

OECD DAC. (1996). *Shaping the 21st Century: The contribution of development co-operation*. Paris: OECD.

OECD DAC. (2006). *The challenge of capacity development: Working towards good practice*. Paris: OECD.

Olson, M. (2000). *Power and prosperity: Outgrowing communist and capitalist dictatorships*. New York: Basic Books.

Ortiz, A. (2013). *Capacity building in complex environments. Seeking meaningful methodology for social change*. Brighton: Institute of Development Studies, University of Sussex.

Pearson, J. (2011a). *The core concept, Part I*. Available at: www.Lencd.org/learning.

Pearson, J. (2011b). *Creative capacity development: Learning to adapt in development practice*. Westhart, CT: Kumarian Press.

Pender, J. (2001). Structural adjustment to comprehensive development framework: Conditionality transformed, *Third World Quarterly, 22*(3), 397–411.

Perroux, F. (1973). *Pouvoir et économie*. Paris: Bordas.

Psacharopoulos, G., & Patrinos, H.A (2002). Returns to investment in education: A further update, policy research Working Paper 2881, Washington, DC: World Bank.

Sagar, A. (2000). Capacity development for the environment: a view for the south, a view for the North. *Annual Review of Energy and Environment. 25*, 377–439.

Samuelson, P. A. (1954). The pure theory of public expenditure. *The Review of Economics and Statistics*, *36*(4), 387–389.

Sen, A. (1997). Editorial: Human capital and human capability. *World Development*, *34*(3), 1959–1961.

Sen, A. (1999). *Development as freedom*. Oxford: Oxford University Press.

Shyamsundar, P. (2002). Poverty and environment indicators. Environmental economic series; no. 84. Washington, DC: World Bank. Available at: http://documents.worldbank.org/curated/en/474441468324048267/Poverty-and-environment-indicators

Stiglitz, J. E. (1999). Knowledge as a global public good. In I. Kaul, I. Grunberg, & M. Stern (eds), *Global public goods*. New York: Oxford University Press.

Swedish Ministry of Foreign Affairs (1997). *Partnership with Africa: Proposals for a new Swedish policy towards Sub-Saharan Africa*, Stockholm: Swedish Ministry of Foreign Affairs.

UNDP. (1993). *National capacity building. Report of the administrator*. DP/1993/23.

UNDP. (1996). *Building sustainable capacity: challenges for the public sector*. New York: United Nations Development Programme.

UNDP. (1997). *Governance for sustainable development*. Geneva: UNDP.

UNDP. (2009a). Capacity development: A UNDP primer. United Nations Development Programme. Available at: www.undp.ro/publicatio/capacitydevelopment

UNDP. (2009b) Reforming technical cooperation for capacity development. Available at: www.case-research.eu/en/reforming-technical-cooperation-for-capacity-development

UNDP and Berg, E. (Coordinator). (1993) *Regional Bureau for Africa: rethinking technical cooperation: reforms for capacity building in Africa*. New York: UNDP.

UNDP-ODI. (2014). Where next for aid? The post-2015 opportunity. Discussion Paper. New York: UNDP.
UNFCCC. (2001). *Report of the Conference of the Parties on its Seventh Session, held at Marrakech from 29 October to 10 November 2001*. FCCC/CP/2001/13/ Add.1. 21 January. Bonn: United Nations Framework Convention on Climate Change.
UNFCCC. (2015). Synthesis report on the aggregate effect of the intended nationally determined contributions. Available at: https://unfccc.int/resource/docs/2015/ cop21/eng/07.pdf
UNFCCC. (2016, April 27). Status of ratification of the Kyoto Protocol. Available at: http://unfccc.int/kyoto_protocol/status_of_ratification/items/2613.php
Victor, D. (2013). Foreign aid for capacity building to address climate change: Insights and application. WIDER Working Paper. No. 2013/084.
Vincent-Lancrin, S. (2007). Developing capacity through cross-border tertiary education. In OECD-World Bank, *Cross-border tertiary education: A way towards capacity development* (pp. 47–102). Paris: OECD.
Winthrop, R., & McGivney, M. (2016). *Why wait 100 years? Building the gap in global education*. Washington, DC: Brookings Institution.
Wohlgemuth, L. (2005). A perspective for the future of Africa with competence building in focus. In M. Beveridge, K. King, R. Palmer, & R. Wedgewood (eds), *Reintegrating education, skills and work in Africa*. Edinburgh: University of Edinburgh.
Wolfensohn, J. D. (1998). The other crisis: Address to the board of governors, World Bank, Washington, DC. October 6, 1998. Available at: https://openknowledge.worldbank.org/handle/10986/26163
Wood, B., Betts, J., Etta, F., Gayfer, J., Kabell, D., Ngwira, N., Sagasti, F., & Samaranayake, M. (2011). *The evaluation of the Paris Declaration, Phase 2: Final report*. Copenhagen: Danish Institute for International Studies.
World Bank. (1998). *Assessing aid: what works, what doesn't and why*. New York: Oxford University Press.
World Bank (1999). A proposal for a comprehensive development framework. Discussion Paper. Washington, DC. World Bank. Available at: http://worldbank.org/cdf/
World Bank. (2005). *Capacity building in Africa: An independent evaluation*. Washington, DC: World Bank.
World Bank. (2009a). Comprehensive Development Framework, a discussion draft. Available at: www.worldbank.org/cdf/
World Bank. (2009b). Defining capacity is complex. Available at: www.lencd.org/ learning. (accessed 19 September 2009).
World Bank. (2011a). *The state of World Bank knowledge services: Knowledge for development*. Washington, DC: World Bank.
World Bank. (2011b). *A new instrument to advance development effectiveness: Program-for-results*, Washington, DC: World Bank.
World Bank. (2014). *Indicators of the Strength of Public Management Systems (ISPMS): Summary of the iChallenge*. Washington DC: World Bank.

5 Case studies

Bangladesh, Uganda and Jamaica

With Shaila Mahmud, Revocatus Twinomuhangi, Joseph Epitu and Stacy-ann Robinson

Introduction

With the capacity-building framework laid out in Chapter 4, we now begin to examine the impacts and vulnerabilities of how such efforts have been carried out on the ground. Three country case studies have been selected – Bangladesh, Uganda and Jamaica. The first two are least developed countries (LDCs), one a land-locked country from Asia and the other a coastal country in Africa. while Jamaica is a small island developing state (SIDS). These countries were selected with the understanding that LDCs and SIDS are impacted most by climate change. These two groups of countries have the highest need for capacity building, but these needs and efforts may differ in important ways. In addition to the UNFCCC synthesis reports on capacity building, there are a number of consultancy-based reports of bilateral and multilateral donor agencies, but very little scholarly literature. This process involves a review and analysis of available country-based studies and reports undertaken by governments, bilateral and multilateral agencies. It has been substantiated by interviews with a cross-section of stakeholders on the supply and demand sides of capacity building in these three case study countries. Based on reviews from these three countries, there are similarities and differences in vulnerabilities, policies, institutional mechanisms and capacity building efforts.

Impacts, vulnerabilities and institutional mechanisms for capacity building in Bangladesh

Despite many climate capacity-building projects by the government, development partners and various NGOs, Bangladesh still suffers from capacity constraints, both in terms of technical knowledge and human resources. These constraints are due mainly to lack of both coordination and long-term and sustainable support from donors. This review of climate change policies in Bangladesh shows that the need for capacity building as a building block to effectively address mitigation and adaptation has not yet been well understood. Policy documents tend to demonstrate theoretical approaches and aspirations rather than actions on the ground.

Impact and vulnerability from climate change

After the city-state of Singapore, Bangladesh is the most densely populated country in the world, with a population of about 160 million and area of 147,570sq.km. However, its population growth rate (less than 1.4 per cent) is comparable to many developed countries. It is a low-lying country located in South Asia with a coastline of about 700km (360miles) on the northern littoral of the Bay of Bengal. The delta plain of the Ganges/Padma, Brahmaputra/Jamuna, and Meghna Rivers (the GBM basin) is one of the most fertile lands on earth, and the tributaries of these rivers occupy almost 80 per cent of the country. Straddling the Tropic of Cancer, Bangladesh has a tropical monsoon climate characterized by heavy seasonal rainfall, high temperatures and very high humidity. About 88 per cent of the country's landmass consists of floodplains, and it sits in a delta with a low elevation of 10metres above mean sea level (GED 2015).

Most of the country is farmed intensively, with rice grown in three seasons as the main crop. The country is also experiencing rapid urbanization, as well as associated industrial and commercial development. For the last quarter century, economic growth rates have ranged between 6 and 7 per cent, and about one-fifth of the population lives below the poverty line, wreaking havoc on the limited natural base of the country. Agriculture now contributes less than one-fifth of the country's GDP, while the export of ready-made garments provides about 80 per cent of the country's total export earnings.

However, the omnipresent threat of climate change already endangers the lives of the marginalized people who are more exposed to climate-related shocks, such as floods and other extreme weather events, due to hydro-geological as well as poor socio-economic conditions (Rozenberg & Hallegatte 2015). The location of Bangladesh at the epicentre of the South Asian monsoon region, coupled with its socio-economic vulnerabilities, makes it extremely vulnerable to the adverse impacts of climate change. Due to its location in the tropics, the dominance of floodplains, low elevation, high population density and low economic and technological capacity, Bangladesh is susceptible to extreme weather events such as cyclones, tidal surge and tidal waves; floods, submergence and water logging, erratic rainfall, increase in temperature and low rainfall and even drought (MoEF 2005; DoE 2007; Huq & Rabbani 2011). Despite contributing less than 0.1 per cent of global greenhouse gas emissions, Bangladesh is likely to be hit hardest by the effects of climate change (Huq 2001). According to the Global Climate Risk Index Report by Germanwatch, Bangladesh has been ranked sixth among the 10 countries worst affected by extreme weather events between 1996 and 2015 (Kreft,

Eckstein, & Melchior 2017). The recent evidence of temperature rise due to human actions indicates more severe consequences in the future (GoB & UNDP 2009). Moreover, the country is predicted to experience an increase in average daytime temperatures of 1 °C by 2030 and of 1.4 °C by 2050 (FAO 2006; IPCC 2007).

The Fifth Assessment Report (AR5) of the Intergovernmental Panel on Climate Change (IPCC) observes the following implications on the future climate change scenario of Bangladesh:

- Warming in Bangladesh is slightly less than the global average due to its location near the ocean and because the tropics generally warm less compared to higher latitudes. Temperatures will increase more in the winter than the summer.
- A prolonged monsoon season will significantly increase precipitation. Rivers will experience increased flows due to more rainfall and melting snow and glaciers in the Himalayas.
- Vulnerability to cyclones is expected to increase. here is also the possibility of higher storm surges, heat waves, floods, prolonged drought, and sea-level rise.

On average, Bangladesh has lost close to 1 per cent of its annual GDP due to extreme weather events during the past 20 years (Kreft, Eckstein, & Melchior 2017). It is projected that climate change could cause an economic loss of 9.4 per cent of its GDP (Ahmed & Suphachalasai 2014). The Climate Change Vulnerability Index (Maplecroft 2011) revealed that climate change would place Bangladesh as one of the worst performing economies worldwide. Climate change will also drastically shift the magnitude of human displacement by the end of twenty-first century (IPCC 2014).

Box 5.1 presents the report on the climate change capacity building in Bangladesh.

Relevant policies in place

Bangladesh was a signatory to the UNFCCC and ratified it in 1994. The country also ratified the Kyoto Protocol in October 2001. Bangladesh submitted its Initial National Communication (INC) to the UNFCCC in November 2002 (MoEF 2002) and its Second National Communication (SNC) in 2012. The SNC presents a comprehensive review of climate change-related policies, plans and strategies up to 2012 for adaptation and mitigation efforts (MoEF 2012). The Government of Bangladesh (GOB), with assistance from the UNDP, is in the process of submitting its Third National Communication (TNC) addressing the post-2012 initiatives related to climate change issues.

Bangladesh was one of the first countries to initiate a series of climate-smart policies and actions, with a focus on adaptation, to address the

Box 5.1 Climate change capacity building in Bangladesh: a case project

The Government of Bangladesh initiated official capacity building by establishing a Climate Change Cell (CCC) in 2004 under the Department of Environment (DoE), with support from the first and second phase of the 'Comprehensive Disaster Management Programme' (CDMP-I and CCP-II) of the UNDP, a multi-donor-supported project. The Asian Development Bank (ADB) under the technical assistance title 'Supporting Implementation of BCCSAP' and 'Climate Change Capacity Building and Knowledge Management' project, prepared a capacity development action plan on climate change and identified priority areas, sectors and cross-cutting areas (climate negotiation, climate financing, legal and economic aspects of climate change, etc.) where capacity building was required immediately. Under the two projects, ADB conducted several short training courses on contemporary aspects of climate change. The Food and Agriculture Organization (FAO) conducted a situation analysis and capacity needs assessment of the Ministry of Environment and Forests (MoEF) and its agencies. The capacity need assessment report focused on building the capacity of the MoEF and its agency in the areas of climate change coordination, ICT, knowledge management, etc. A GIZ-supported project 'Strengthening Climate Finance Governance in Bangladesh' also contributed to capacity building in climate finance. The UK Department for International Development (DFID) have also supported setting up the International Centre for Climate Change and Development (ICCCAD), housed at the Independent University of Bangladesh (IUB), to build capacity towards adaptation to climate change by conducting short training courses and running academic courses.

There is an active LCG-subgroup on climate change capacity building at the MoEF. This subgroup identified nine functional areas on climate change capacity building that includes: (1) climate financing; (2) technical knowledge; (3) project development and management; (4) policy support; (5) coordination capacity; (6) external representations/relations/diplomacy; (7) knowledge management; (8) communications and advocacy; and (9) institutional strengthening and governance. A national-level climate change capacity building committee has been established at the MoEF in line with the Paris Agreement on climate change. Such initiatives have significantly helped Bangladesh in building capacity on climate change.

Certain NGOs, (CARE, Oxfam, ActionAid, IUCN, BCAS, CNRS, etc.), universities (BUET, IUB, BRAC, Dhaka, Chittagong, etc.), research organizations (CEGIS, IWM, BIDS, etc.) and networks (ARCAB, CANSA, etc.) have also conducted short training courses on various aspects of climate change. Some of the universities recently started Master's programmes on climate change, disaster management, in the country. Recently, Dhaka University, Patuakhali University of Science and Technology and BRAC University started Master's programmes on disaster management. NSU has run its undergraduate and graduate programmes in Environmental Science and Management for more than a decade. Under IUB, ICCCAD started a Master's programme on climate change and development, the first of its kind in

Bangladesh. The Bangladesh Public Administration Training Centre (BPATC) regularly conducts a specialized training course on 'Environment and Sustainable Development' for civil servants. They also organize a special course on 'Environment and Disaster Management'. These course modules include sessions on climate change adaptation and disaster risk management. Recently, under the '*National capacity development for implementing Rio Conventions through environmental governance project*', the DoE and UNDP are building capacity of civil servants on the Rio Conventions through engaging public training institutes (e.g. BPATC, BIAM, NATA, NAPD, BCS Administration Academy, etc.). Recently they jointly organized several special training courses with support from UNICEF, UNDP, JICA, KOICA, Australian National University, etc.

Key lessons and policy recommendations

Key lessons and recommendations on climate change capacity building include:

1 Capacity building on climate change should be continued in the relevant ministries and line agencies with a phased implementation approach, appropriate institutional arrangement and tracking mechanism (with baseline, indicators and target) for successful integration of climate change in development planning, programming and budgeting.
2 A continuous flow of support is required for capacity building, institutional strengthening and knowledge management to promote climate-resilient development and green growth.
3 For transformational capacity building programmes on climate change, both the public and private sectors should work together including the engagement of public/private training institutes (e.g. BPATC, NAEM, ICCCAD) to develop and transform individual capacity into institutional capacity with a focus on sustainability, and to scale up activities through use of smart ICT tools and innovative mentoring processes.
4 Immediate, mid-term and long-term capacity building plans with appropriate institutional arrangements and effective coordination and tracking mechanism are required to implement transformational capacity building programmes.
5 Universities produce experts, future leaders or champions on climate change and hence both public and private universities should take the lead on long-term capacity building.
6 Special courses on climate change need to be included/tailored in the training programmes of public service cadres.
7 Instead of stand-alone capacity building projects, components of capacity building should be included in all climate change or relevant development projects and programmes to make a transformational change.
8 North–South, South–South and triangular cooperation will be instrumental to enhance current and future initiatives on capacity building on climate change.

Source: Arif Faisal, UNDP, Dhaka

inevitable consequences of climate change (Huq & Khan 2017). The key guiding policies that promote capacity-building efforts in particular as a means to combat climate change are as follows:

National Adaptation Programme of Action

Based on the guidelines of the UNFCCC LDC Expert Group, the Ministry of Environment and Forests (MoEF) developed and submitted the National Adaptation Programme of Action (NAPA) in 2005, with support from the Least Developed Countries Fund (LDCF). Using a participatory consultation process, the NAPA presented 15 priority areas. These areas included capacity-building efforts, attempts to raise risk awareness and enhance resilience of infrastructure, and implementing mechanisms, and they focused on fisheries, agriculture, water resources, disaster management, insurance, health and industry (MoEF 2005a).

National Capacity Self-Assessment

In 2007, the government undertook the National Capacity Self-Assessment (NCSA) initiative to assess capacity needs and prepare the Bangladesh Capacity Development Action Plan for sustainable environmental governance. The NSCA is the first climate capacity development framework that captures the needs to reinforce the country's capacity to negotiate and implement the global environmental conventions along with strengthening the institutional capacity to produce a comprehensive framework and action plan (MoEF 2007).

The Bangladesh Climate Change Strategy and Action Plan

The Bangladesh Climate Change Strategy and Action Plan (BCCSAP) prepared in 2009 (first one of the LDCs) has been the key guiding document on climate change. It underscored the need for capacity building to make climate change adaptation and mitigation efforts an integral part of the overall development process. The six pillars of BCCSAP address the need to strengthen institutional capacity, build resilience among different groups of people (particularly women and children), and raising risk awareness and capacity for knowledge management (MoEF 2009). In light of the BCCSAP, the Bangladesh Climate Change Trust Fund Act 2010 was enacted by the Parliament to fast-track use of climate funds raised domestically for projects. The ten-year programme under BCCSAP intends to build country capacity and resilience to meet climate change challenges over the next few decades, and to integrate climate change actions into sustainable development strategies.

National Adaptation Plan roadmap

Adopted at COP16 in 2010, the Cancun Adaptation Framework introduced the National Adaptation Plan process as an agency to underline medium- and long-term adaptation needs and to develop and implement strategies and programmes to address those needs with guidance from UNEP and UNDP (UNEP; GED 2015). In Bangladesh, the government finalized the National Adaptation Plan roadmap in 2015.

Nationally Determined Contribution (NDC)

The Nationally Determined Contribution (NDC), known as the Intended NDC (INDC) prior to ratification of the Paris Agreement, is an overarching policy strategy to achieve lower greenhouse gas (GHG) emissions and climate-resilient development. Subject to support received from developed countries, Bangladesh committed to reduce GHG emissions in the power, industry and transport sectors by 5 per cent 'unconditionally' below 'business-as-usual' emissions by 2030, or by a 'conditional' 15 per cent below 'business-as-usual' emissions within 2030. To achieve this goal, the INDC focuses on an implementation roadmap with sets of sectoral action plans, and building capacity based on need. The roadmap suggests the need for good governance and coordination arrangements; setting up institutional arrangements for Measurement, Reporting and Verification (MRV); mitigation efforts through low-emission technologies and efficiencies in power, transport and industry sectors; adaptation measures to mainstream climate change into planning; and regular flow of finance. In this plan, capacity building pursuits will tie together all the needs to implement the NDC.

The 7th Five-Year Plan including climate change as an integral part

The General Economic Division (GED) of the Ministry of Planning has also been active in addressing the need for climate education, institutionalization of climate change, and identification of priority options for adaptation and mitigation. The GED has done so through the country's 7th Five Year Plan (2015), the Perspective Plan for Bangladesh Vision 2021, and the National Sustainable Development Strategy (2010–2021). The Climate Fiscal Framework was also put in place by the Finance Division of the Ministry of Finance (MoF) to provide necessary guidelines for tracking climate-related expenditures, estimating potential costs of long-term financing needs, and identifying institutional weaknesses and skill gaps in enhancing the capacity of the Planning Commission and the Finance Division (MoF 2014).

Institutional mechanisms

In the late 1980s, Bangladesh laid out the institutional framework for environmental management, and later, for addressing climate change. The MOEF, established in 1989, is designated as the central body for carrying out overall activities on climate change, including international negotiations. The Department of Environment and the Department of Forests under the MoEF are also involved in carrying out different climate-related projects. The Climate Change Unit (CCU) was set up as a wing under the MoEF to support inter-ministerial coordination and administration on adaptation and mitigation projects undertaken by the government.

The government has set up two climate change funds to support the implementation of climate change relevant projects: the Bangladesh Climate Change Trust Fund (BCCTF) with domestic budgetary resources, and the Bangladesh Climate Change Resilience Fund (BCCRF) with international support. To date, the BCCTF has disbursed over US$400 million for more than 200 projects envisaged under the BCCSAP (GED 2015). The BCCRF was created with bilateral development partners as a multi-donor fund to streamline the implementation of BCCSAP. The Climate Change Trust, a technical wing under the MoEF, has been established to ensure effective implementation process for the BCCTF projects. Over $200 million from the BCCRF has been allocated to a number of projects to address climate change impacts, primarily related to infrastructure.

The Ministry of Finance and the Ministry of Planning have also been involved in addressing climate change issues for the past decade. The Ministry of Finance has developed a Climate Fiscal Framework (MoF 2014), and the Ministry of Planning ensures that all development projects consider climate change issues in their design and implementation. The External Resources Department (ERD) of the MoF is serving as the National Designated Authority (NDA) for the GCF and to carry on related procedures, including identifying national implementing entities (NIEs) and facilitating stakeholders to locate fundable projects with appropriate application processes for seeking GCF funding.

The MoF has been a key government institution working to integrate actions related to capacity building by introducing risk reduction programmes to enhance human capacity. The various DRR programmes of the Ministry of Disaster Management and Relief (MoDMR) are designed specifically to keep the disadvantaged safe from the effects of natural, environmental and human induced hazards (at a manageable and acceptable level) and to have in place an efficient emergency response system.

The Bangladesh Meteorological Department (BMD) and the Space and Remote Sensing Agency (SPARRSO) (under the Ministry of Defence), and the Flood Forecasting and Early Warning Centre and Bangladesh Water Development Board (under the Ministry of Water Resources), are also engaged in communicating climate capacity-building efforts through various work programmes. Other major development ministries, such as Agriculture, Water and Women and Children Affairs, are also involved in making climate related issues normally part of the sectoral plans. To address the gender knowledge gap, the Ministry of Women and Children Affairs (MOWCA) has an initiative under the BCCTF to support other ministries in integrating gender sensitivity in their respective project designs (Ahmed et al. 2015).

Capacity-building initiatives

Disaster-prone Bangladesh has been able to learn from the variety of approaches it has used to fight climate change over the years. Development partners and NGOs/CSOs, in consultation with concerned communities, have stepped up building resilience through the plans and programmes they have adopted. One ground-breaking initiative of the government is the Cyclone Preparedness Programme (CPP). The programme is considered a model programme in the disaster management field (MoDMR 2016). Other international and national organizations and departments involved include the Comprehensive Disaster Management Programme under UNDP, Save the Children, World Bank, UNICEF, etc.

A number of donors, including the UNDP, UKAid, AusAid, DFID, EU, Norwegian Embassy, SIDA, the Bangladesh Red Crescent Society, Oxfam GB, CARE Bangladesh, ActionAid, IUCN, and others have long worked on disaster management in Bangladesh. Their projects have included community disaster preparedness and community-based development and capacity-building initiatives.

One of the key civil society initiatives in implementing long-term capacity building in research is the Gobeshona (Research) Platform (Gobeshona n.d.). The Platform was coordinated by the International Centre for Climate Change and Development (ICCCAD) and established at the Independent University of Bangladesh (IUB). It serves as a knowledge and research hub to help involve stakeholders such as youth in addressing climate change (www.gobeshona.net).

Again, despite the many climate capacity-building projects by the government, development partners and various NGOs, Bangladesh still suffers from capacity constraints, both in terms of technical knowledge and human resources. These constraints are due mainly to lack of both coordination and long-term and sustainable support from donors. About 60 per cent of the BCCTF has been used to finance over 300 projects on climate change that are adaptation-focused (rather than focused on mitigation) (Ahmed et al. 2015). However, only a negligible 3 per cent of the BCCTF has been dedicated to capacity building and institutional strengthening. The Asian Development Bank, (ADB) under the technical assistance projects 'Supporting the Implementation of BCCSAP' and 'Climate Change Capacity Building and Knowledge Management' drafted a capacity development action plan on climate change and identified priority areas, sectors, and cross-cutting areas (climate negotiations, climate financing, legal and economic aspects of climate change, etc.) where capacity building is required.

The MoEF recently concluded a project entitled, 'Strengthening the Environment, Forestry and Climate Change Capacities of the MoEF and its Agencies.' The project was funded by USAID and had technical assistance from FAO, and developed a Country Investment Plan (CIP) for the Environment, Forestry and Climate Change sector (MoEF 2016). The capacity need assessment report drafted by FAO focused on building the capacity of MoEF and its agency in the areas of climate change coordination, ICT, knowledge management, etc.

Often, the inadequacy of monitoring and evaluation within different agencies leads to irregularities in reporting and impairs the right to access information. In response to these setbacks, ERD has generated the Aid Information Management System (AIMS) as part of the GoB's transparency initiative. The Climate Finance Transparency Mechanism (CFTM) project is also being implemented by the International Centre for Climate Change and Development (ICCCAD), the Bangladesh Centre for Advanced Studies and the Centre for Climate Change and Environmental Research under BRAC University in partnership with ERD under the AIMS. This project aims to improve the overall levels of transparency in governance and to increase the efficacy of climate change funds from the government and from donors in tackling the impacts of climate change in Bangladesh.

We have seen that although Bangladesh has undertaken several policy responses to climate change, there is not yet a detailed understanding of the impacts in a way that will allow words to be translated into actions. As suggested at the outset, this review of climate change policies in Bangladesh shows that the need for capacity building as a building block to effectively address mitigation and adaptation is not yet understood. Policy documents still tend to demonstrate theoretical approaches and aspirations, rather than actions on the ground.

Uganda case study

This case study of Uganda shows that human resource capacity continues to limit the implementation of climate change plans and policies. Capacity gaps in climate information and knowledge management restrict Uganda's ability to manage climate risks and increase resilience. For example, inadequate meteorological observation networks, systems and staffing decrease the UNMA's ability to make weather forecasts or to monitor, detect and predict climate variability and climate change (World Bank 2015b). Uganda lacks technical expertise in climate change modelling and inventory management, and many technical staff and decision-makers lack basic awareness of climate change. The sharing of data and information is also challenging. While several institutions have developed databases, they are rarely shared to inform planning. Weak inter-institutional collaboration among climate change actors is a challenge for Uganda; uncoordinated projects and activities have led to duplicate or overlapping interventions across sectors, civil society and development partners. This lack of coordination hinders an integrated approach to climate change and disaster management, especially at the district level. Limited opportunities for capacity building and training of technical staff will continue to hinder human capacity development.

Climate impacts and vulnerabilities

Uganda is a landlocked country in East Africa with a total area of 241,550 km^2. Of this, 41,743 km^2 (17.2 per cent) is open water and swamps, and 199,807 km^2 is terrestrial (Government of Uganda 2015a). On average, Uganda's land area is 1200 metres above sea level, with its lowest point (in the Albert Nile) at 620 metres and its highest at 5100 metres (the peak of Mt. Rwenzori) (Government of Uganda 2014a). Uganda's various lakes and rivers make it rich in water resources. The country's most significant river is the Nile River, one of the world's longest rivers, and whose source is Lake Victoria, the largest lake in Africa (Twinomuhangi et al. 2014). Some 80 per cent of Uganda's approximately 35 million people live in rural areas, while only 20 per cent reside in urban areas. Uganda has one of the world's fastest-growing populations, with an annual population growth rate of 3 per cent (Government of Uganda 2016).

Uganda is a Least Developed Country (LDC), and has a per capita GDP of US$623 (World Bank 2016). Agriculture is the country's main economic activity, and supports the livelihood of 73 per cent of households. It employs about 34 per cent of the country's economically active population, and over 80 per cent of its poorest population (Government of Uganda 2015b). The population living below the poverty line dropped from 56 per cent in 1992 to 24.5 per cent in 2013 (Government of Uganda 2014b), but Uganda is still unlikely to meet its poverty eradication objective of reducing absolute poverty to below 10 per cent by 2017 (Amone 2014).

Uganda has a mainly tropical climate and bimodal rainfall distribution. The main rainy seasons are from March to May and October to December, though the northern part of the country has a single rainy season from March to mid-October. Mean daily temperatures are 28 °C and the long-term mean near-surface temperature is 21 °C. The highest temperatures are observed in the north, especially the north-east, while lower temperatures occur in the south (Government of Uganda 2009, 2015b; UNDP, 2013). These mild conditions make climate one of Uganda's most valuable natural resources (Government of Uganda 2015b).

Historic trends show that Uganda's climate is variable and changing. Between 1960 and 2000, the country's average annual temperatures increased by 1.3 °C (Government of Uganda 2015b; World Bank 2015a). During the same period, the frequency of hot days increased by 20 per cent and the frequency of cold days decreased across all months, except December, January and February (McSweeny et al. 2010). The country has not experienced any statistically significant changes in annual rainfall over the past 60 years (World Bank 2015a), but a USAID (2013) study reported changes in the onset of rainy seasons, with rains beginning 15–30 days before or after the usual season onset, shortening the seasonal length by 20–40 days. The Future Climate for Africa (2016) reports a significant decrease in rainfall experienced in northern Uganda, and a reduction in rainfall during the March–May rainy season at rate of 6 mm per month per decade (McSweeny et al. 2010). A significant increase in heavy rainfall events has also been reported.

Floods and droughts, the most common climate hazards in Uganda, have been on the rise over the past 25 years. Severe droughts occurred in 2001, 2002, 2005, 2008 and 2011, and between 1991 and 2000 there were seven droughts in the Karamoja sub-region alone. Floods are also common in eastern Uganda and Kampala city (Government of Uganda 2014a, 2014b).

Climate projections for Uganda indicate increases in near-surface temperatures by 2 °C and 3 °C in next 50 years (Ministry of Water and Environment 2015). By the 2060s, 'hot' days are projected to make up between 15–43 per cent of the year (Future Climate for Africa 2016). Extreme events (mainly droughts, extreme rainfall, floods and heatwaves) are also expected to increase in the future.

Annual rainfall totals are not projected to change greatly, remaining within a range of plus or minus 10 per cent from the present. However, rainfall totals may drop significantly over Lake Victoria (-20 per cent from the present) and some models suggest a 14 per cent increase in heavy rain events by 2060 that would increase run-off and floods in many parts of the country (McSweeny et al. 2010; Ministry of Water and Environment 2014; USAID 2015). Climate change is also causing ice to melt on Mt. Rwenzori, where the surface ice area decreased by 49 per cent between 1955 and 2003 and is projected to disappear within the next two decades (Government of Uganda 2014a).

Uganda is ranked as the country 27th most vulnerable and 25th least prepared to address climate change on the ND-GAIN Index (Notre Dame Global Adaptation Initiative 2017). Various climate change vulnerability studies (Government of Uganda 2002, 2007, 2014a; UNDP 2013; USAID 2013; FAO 2015; Ministry of Water and Environment 2015) report that the country is highly vulnerable to the impacts of climate change, and that this high vulnerability is mainly due to the dependence of Uganda's economy and population on climate-sensitive natural resources and sectors

– agriculture, energy, transport, tourism, water, human health and nutrition, education (World Bank 2015b). This vulnerability is exacerbated by the high poverty levels, high population growth and poor stewardship of the natural resource base that weakens resilience (Government of Uganda 2014b, 2017). Agriculture accounts for 22 per cent of GDP, 85 per cent of export earnings and 68 per cent of total employment (Uganda Bureau of Statistics 2014). Uganda's agriculture is largely rain-fed and subsistence-oriented, and so is highly vulnerable to rainfall patterns and droughts. Irrigated agriculture comprises only 0.1 per cent of the total cultivated land (FAO 2012; OECD 2012; World Bank 2012a). Over the last decade, crop productivity growth has been on a downward trend due to low technology development, averaging around 1 per cent per year (UNDP 2015). For example, fertilizer is used at one of the lowest rates in the world (an average of 1 kg of nutrients per hectare), and only 6.3 per cent of farmers use improved seeds (Hundsbæk et al. 2012). Food insecurity is on the rise, especially in the northern region of the country. In the Karamoja sub-region, about 30 per cent of food needs are covered by aid (USAID 2011).

Climate change also has a major impact on coffee, Uganda's leading export crop. By 2050, the value of coffee could fall by 50 per cent due to a decrease in the area suitable for its production. This could cause losses up to an estimated US$1.24 million (Twinomuhangi & Monkhouse 2015).

Uganda's existing infrastructure is in poor condition, increasing the chances of destruction and damage from climate extremes like floods, storms and landslides (World Bank 2015b). The country depends heavily on road transport, which is highly susceptible to damages from storms, floods and landslides. The recent reduction in water levels due to drought has also affected the electricity generation by Uganda's major power dams (Nalulabale and Mpanga). By 2050, water availability may have declined by an estimated 26 per cent (Twinomuhangi & Monkhouse 2015). As Uganda's temperature and rainfall extremes increase, cases of malaria and waterborne diseases like cholera and dysentery will also rise (World Bank 2015b).

Policies in place

Uganda is a Party to the UNFCCC, which it signed and ratified in 1992 and 1993 respectively. Uganda has also ratified the Paris Climate Change Agreement. The country developed and submitted its initial and second National Communications to the UNFCCC in 2002 and 2014, comprising a national GHG inventory, vulnerability and adaptation to climate change, and recommendations for responding to climate change. Uganda prepared and submitted its INDC (now NDC) to the UNFCCC, which details Uganda's contribution to taking on climate change.

Article 39 of Uganda's Constitution (1995) guarantees the right of every person to a clean and healthy environment. The Constitution advocates for the protection and preservation of the environment, management of the environment for sustainable development, and promotion of environmental awareness. The Uganda Vision 2040 articulates Uganda's development agenda and goals to transition from a low- to upper-middle-income status by 2040 (Government of Uganda 2010). Through the Vision 2040, Uganda hopes to achieve a green economy, a clean environment and climate change resilience in the context of sustainable development and poverty eradication. Uganda's five-year National Development Plan (NDP) 2015/2016 to 2019/2020 incorporates climate change as a cross-cutting issue and aligns with the global Sustainable Development Goals (Government of Uganda 2015d).

Uganda's climate change policy framework is guided by the National Climate Change Policy of 2015 (NCCP), which aims to harmonize and coordinate a climate-resilient and low-carbon development path for sustainable development in Uganda (Government of Uganda 2015e). The policy focuses on both adaptation and mitigation, and emphasizes climate change mainstreaming across all sectors, capacity development and access to climate finance. Uganda is also developing a Climate Change Law to provide an overarching legal framework to guide climate change coordination, implementation and reporting.

Uganda has additional sectoral policies that enhance climate change action. Its meteorology policy aims to promote awareness, monitor weather and climate, maintain a climate database, and provide regular advice on the state of weather and climate. A Disaster Preparedness and Management (DPM) Policy was put in place in 2010 to establish institutions and mechanisms to reduce the vulnerability of people, livestock, plants and wildlife to disasters. The policy focuses on risks, including those related to climate hazards (especially droughts, floods and landslides). The National Agricultural Policy 2013 focuses on supporting sustainable agricultural production, food security, increased household incomes and the ways to achieve these goals through promoting climate-smart agriculture.

Several of the country's policies were developed before climate change became a major concern, and so do not adequately address climate issues. The Energy Policy of 2002 aims to meet the energy needs of the Ugandan population for social and economic development in an environmentally sustainable manner. The Renewable Energy Policy of 2007 aims to increase the use of modern renewable energy from 4 per cent to 61 per cent of total energy consumption in the country by 2017. With support from the EU/FAO Global Climate Change Alliance (GCCA) project, the Ministry of Agriculture, Animal Industry and Fisheries has developed an agricultural sector National Adaptation Plan.

Institutional mechanisms

Uganda has put a climate change institutional framework in place to spearhead response to climate change, and the country's Ministry of Water and Environment (MWE) is required to lead climate policy and implementation. The Ministry has a Climate Change Department (CCD) that acts as the UNFCCC Focal Point. In line with the NCCP, the CCD may be elevated to a semi-autonomous agency like the Climate Change Commission. Currently, the CCD faces enormous capacity challenges, coming up against institutional, staffing and finance constraints in particular. As a Department of the Ministry, one challenge for the CCD is in influencing other powerful Ministries, Departments and Agencies (MDAs), for instance, the Ministry of Finance, Planning and Economic Development (MoFPED), the Office of the Prime Minister (OPM), and the National Planning Authority (NPA) among others. The CCD approved staff structure still has 40 per cent of its positions unfilled.

As an LDC, Uganda cannot mobilize adequate climate change finance and requires international support. Most importantly, it needs access to the Green Climate Fund (GCF) and the Adaptation Fund (AF). Over the last decade, the CCD has been receiving most of its funding (70 per cent of its operational budget) from DANIDA through the Joint Water and Environment Sector Programme Support Programme. However, DANIDA funding for the programme will end in June 2018, restricting the CCD's abilities (Ministry of Water and Environment 2017). Uganda is not yet accredited to the GCF and AF, though the National Implementing Entity (NIE) has been identified as the MWE and the Designated National Authority (DNA) is MoFPED. While it is important for Uganda to access climate financing streams to support development, inadequate human capacity to develop bankable projects must also be addressed (World Bank 2015b).

A National Climate Change Policy Committee (NCCPC), chaired by the Prime Minister brings together ministers from the various MDAs and oversees climate change policy formulation and implementation. A multi-sectoral National Climate Change Advisory Committee (NCCAC), chaired by the Minister of MWE is in place to ensure working-level coordination across MDAs and to provide technical inputs to the NCCPC. In addition to technical representation from MDAs, the Advisory Committee has representatives from private-sector associations, civil society, academia and local government. Both the NDP and NCCP require MDAs to mainstream climate change in sectoral plans and budgets. The National Planning Authority (NPA) is responsible for ensuring that all sectoral plans and District Development Plans (DDPs) integrate climate change, and all sectors have designated people to act as climate change focal leads. At the local level, climate change focal points are anchored within the District's Natural Resources Departments. The Ministry of Local Governments

(MoLG) provides guidance to Districts to integrate climate change into DDPs and budgets. However, human resource capacity continues to limit implementation of climate change plans and policies. Many technical staff and decision-makers lack basic awareness on climate change. In particular, Uganda lacks technical expertise in climate change modelling and inventory management. The limited opportunities for capacity building and training of technical staff will continue to hinder human capacity development.

The Uganda National Meteorology Authority (UNMA) is the agency responsible for collecting and disseminating weather and climate information. The UNMA is also the IPCC focal point, and works with the WMO to fulfil its mandate. Capacity gaps in climate information and knowledge management restrict Uganda's ability to manage climate risks and increase resilience. For example, inadequate meteorological observation networks, systems and staffing decrease the UNMA's ability to make weather forecasts or to monitor, detect and predict climate variability and climate change (ibid.). Sharing data is also challenging. While several institutions have developed databases, they are rarely shared to inform planning.

To improve climate change reporting, Uganda launched a national GHG inventory management system in October of 2016 (with support from the EU/UNDP Low Emission Capacity Development project) (UNDP 2016). However, the country still faces serious constraints in reporting on climate change. Data availability, access and formats restrict the compilation of GHG inventories, especially for national communications. In most cases, data is lacking or in a difficult form or quality to process. Uganda does not have a functional Measurement Reporting and Verification (MRV) system to cover adaptation, mitigation and financing. Most MRV activities are project-based rather than institutionalized, and baselines and indicators are lacking in many cases (ACCRA 2017). A national Performance Measurement Framework (PMF) has been developed to track the implementation of the NCCP, but has not yet been operationalized.

Again, a key challenge for Uganda is weak inter-institutional collaboration among climate change actors. Uncoordinated climate change projects and activities have led to duplicated or overlapping interventions across sectors, civil society and development partners. For example, though climate risk management and disaster preparedness should complement one another, there is a division of responsibility in coordinating the two. The Department of Disaster Management and Refugees at the OPM is responsible for disaster risk management while the CCD oversees climate change, and there is limited cooperation and exchange of information between these two coordinating departments (World Bank 2015b). This lack of coordination hinders an integrated approach to climate change and disaster management, especially at the district level.

Civil society is active on climate change in Uganda. International NGOs working in Uganda include the WWF, Care International, Oxfam, Save the Children-Uganda, and World Vision International. The latter four have formed a consortium called the Africa Climate Change Resilience Alliance (ACCRA) (Government of Uganda 2017). Local NGOs have formed the Climate Action Network Uganda (CAN-U), an umbrella organization to collectively address climate change issues.

Jamaica case study

Climate impacts and vulnerabilities

Jamaica is a small island state in the Caribbean Sea, located just south of Cuba and the Cayman Islands, and close to halfway between the southernmost tip of the US State of Florida and the Panama Canal. The country has a population of 2.7 million, and tourism is its main source of foreign exchange. Cold fronts often occur between October and April, and other tropical weather systems (e.g. tropical depressions, waves, hurricanes and storms) between April and December (Meteorological Service of Jamaica 2002b). The hurricane season runs from June to November (ibid.).

Over the last 30 years, Atlantic hurricanes and tropical storms affecting Jamaica have increased in intensity (CARIBSAVE Partnership 2012). Their intensity will continue to increase, though they may decrease in frequency due to lower vertical wind shear in a warmer climate (McCalla 2012). Increases in nearby sea surface temperatures (SSTs) have also been observed, and General Circulation Model (GCM) projections show an increase in annual mean SSTs of between 0.9 and 2.7 °C relative to the 1970–1999 average by the 2080s (CARIBSAVE Partnership 2012). These observed and projected SST increases suggest that increases in hurricane and storm activity will continue (ibid.).

Regional Climate Model (RCM) simulations show that under a high emissions scenario, annual mean temperatures in Jamaica will increase between 2.9 and 3.4 °C by the 2080s (ibid.). It is also likely that Jamaica will experience heat waves (McCalla 2012). Both RCM and GCM simulations show annual rainfall decreasing between 10 and 41 per cent, especially between the months of March and August (CARIBSAVE Partnership 2012). Additionally, sea levels are expected to rise between 0.21 and 0.48 m under an A1B scenario (i.e. where GHG emissions are balanced across all sources) by 2100 (McCalla 2012). Sea level rise is also expected to cause increased coastal erosion and inundation, saltwater intrusion, flooding and the inward retreat of estuaries.

The Vulnerability and Assessment Chapter in Jamaica's Initial National Communication to the UNFCCC (submitted in November 2000) provided an initial assessment of the country's vulnerability to climate change (Government of Jamaica 2000). Consultations and interviews with key stakeholders in Government Ministries, Agencies and Departments, NGOs and the private sector, along with desk-based research, identified the coastal zone, water resources and agriculture as among the most vulnerable sectors (ibid.). The country's Second National Communication to the UNFCCC, submitted in December 2011, contained more in-depth assessments of vulnerability and adaptation for five sectors: (1) the coastal zone and human settlements; (2) water resources; (3) agriculture; (4) tourism; and (5) human health (Government of Jamaica 2011). These later assessments reiterated the increasing vulnerability of Jamaica's coastal and water resources to the impacts of climate change. Human settlements in the coastal zone have a combined population of almost 680,000 people who are directly connected to the coastline and are at greater risk of coastal hazards (Lyew-Ayee Jr 2015). Increasing exposure and sensitivity to coastal hazards will have negative impacts on the population's economic and social well-being (Government of Jamaica 2011; McCalla 2012).

Over half of the country's economic assets, such as air and sea ports, are located in coastal areas (Richards 2008). Coral reefs, which are integral to the island's ecosystem, are vulnerable to SST increases and coral bleaching, which in turn negatively impact reef fauna and increase coastal vulnerability to erosion (Government of Jamaica 2000, 2011). Tourism is a key economic activity, and depends on coastal infrastructure and coastal attractiveness (McCalla 2012). In 2013, travel and tourism generated a total earning of US$4 billion, which accounted for 28.5 per cent of Jamaica's GDP (considering its direct, indirect and induced impact on GDP) (World Travel & Tourism Council 2013).

In terms of the water resources sector, groundwater meets roughly 80 per cent of Jamaica's water demands and is 84 per cent of the country's exploitable water supply (Government of Jamaica 2011). Data on stream flows from the Government's Water Resources Authority shows that several rivers are under stress and others are in deficit, leading to water lock-offs and limiting the country's overall water supply (ibid.). With annual rainfall projected to decrease between 10 and 41 per cent, Jamaica may not be able to meet its demand for water in the near future (Government of Jamaica 2011; CARIBSAVE Partnership 2012).

Relevant policies in place

McCalla (2012) provides an excellent comprehensive review of Jamaica's policies, guidelines, plans, legislation and regulations that support and advance the country's climate mitigation and adaptation agendas. The review identified 29 draft and in-force policies and guidelines, 8 plans, and 30 pieces of legislation and regulations. At a minimum, these instruments cover some aspect of adaptation and/or mitigation in over ten sectors, including agriculture and food security, housing and human settlements, fisheries and forestry, water, disaster resilience and tourism (ibid.). Examples of the relevant policies/guidelines and plans include:

1 Vision 2030 Jamaica: National Development Plan
2 National Climate Change Policy and Action Plan
3 National Energy Policy 2009–2030
4 National Environment Action Plan
5 Water Sector Policy, Strategies and Action Plan
6 Natural Resources Conservation Authority Guidelines for the Planning, Construction and Maintenance of Facilities for Enhancement and Protection of Shorelines
7 National Forest Policy.

Of particular importance are the first two policies and plans listed above. Vision 2030 Jamaica, which was approved in 2009, is 'a strategic roadmap to guide the country to achieve its goals of sustainable development and prosperity by 2030' (Planning Institute of Jamaica 2009; Jamaica Information Service 2017). It 'offers a comprehensive planning framework in which the economic, social, environmental and governance aspects of national development are integrated' and aims to 'put Jamaica in a position to achieve developed country status by 2030 and in the process, make it "the place of choice to live, work, raise families, and do business"' (Nachmany et al. 2015: 3). The Plan has seven guiding principles, including sustainability (environmental, social and economic), and four National Development Goals, each with associated 15 National Outcomes. Outcome 14 tackles hazard risk reduction and climate change adaptation (Planning Institute of Jamaica 2017). Under this National Outcome, two overarching National Strategies are to be pursued (Nachmany et al. 2015; Grey 2017). The first is developing climate change adaptation measures and the second is contributing to slowing global climatic change (ibid.). The Planning Institute of Jamaica (PIOJ) is responsible for carrying out Vision 2030 Jamaica (Planning Institute of Jamaica 2012).

The National Climate Change Policy Framework and Action Plan was tabled as a Green Paper in the Parliament of Jamaica in 2013 (Nachmany et al. 2015). It has three aims (Government of Jamaica 2013: 9):

1 Mainstreaming climate change considerations into sectoral and financial planning and build the capacity of sectors to develop and implement their own climate change adaptation and mitigation plans.
2 Supporting the institutions responsible for research and data collection at the national level on climate change impacts to Jamaica to improve decision-making and prioritization of sectoral action planning.
3 Improving communication of climate change impacts so that decision-makers and the general public will be better informed.

Developed by the then Ministry of Water, Land, Environment and Climate Change with financial support from the United Nations Environment

Programme, the United States Agency for International Development, and the European Union, the Green Paper benefitted from island-wide public consultations in 2014 (Nachmany et al. 2015; Williams-Raynor 2015c). It outlines existing and new initiatives that are priorities for implementation (Government of Jamaica 2013). For the Climate Change Division (CCD) in the Ministry of Economic Growth and Job Creation, which coordinates Jamaica's involvement in national, regional and international climate actions, this means aligning climate actions with the Medium-Term Socio-Economic Framework and with recommendations from stakeholder consultations (Government of Jamaica 2013, 2016). In March 2015, the Cabinet approved the Draft Climate Change Policy (Williams-Raynor 2015c).

Institutional mechanisms

Before the CCD was established, the Meteorological Service of Jamaica was responsible for coordinating Jamaica's response to climate change. In particular, it was responsible for collecting and sharing data on climate impacts and vulnerabilities, providing technical support to the government in international climate negotiations, and coordinating the preparation of the country's National Communications. In 2015, the government unveiled plans to merge the Meteorological Service with the CCD to create the National Agency for Meteorology and Climate Resilience (Williams-Raynor 2015a). With the change in political administration in 2016, however, the Meteorological Service became a Department of the Ministry of Economic Growth and Job Creation, and the CCD a Division of the same Ministry (Ministry of Economic Growth and Job Creation 2017; Office of the Prime Minister 2017).

The Office of Disaster Preparedness and Emergency Management (ODPEM) is an Agency of the Ministry of Local Government and Community Development, and is responsible for disaster management in Jamaica (Ministry of Local Government and Community Development 2013). The ODPEM takes 'action to reduce the impact of disasters and emergencies on the Jamaican population and its economy', irrespective of whether these disasters and emergencies are climate change-related or not (Office of Disaster Preparedness and Emergency Management 2008e). ODPEM works closely with a number of other MADs on disaster preparation and management, including the Meteorological Service of Jamaica and the Jamaica Information Service. The Caribbean Disaster Emergency Management Agency is another key partner.

Local NGOs in Jamaica do not play a lead role in collecting and sharing data on climate impacts and vulnerabilities. Rather, many environmental NGOs focus on education and raising awareness. For example, PANOS Caribbean, a regional NGO, operates in Jamaica with goals to 'enable the people of the Caribbean to conceive, drive and communicate their own development agenda' and to 'develop media, information and communication partnership communicate towards development' (PANOS Caribbean 2016b). This involves sharing climate-related data with citizens and using data to develop communication campaigns. The Environmental Foundation of Jamaica, another NGO, funds community-based capacity-building projects and its 2013–2015 Strategic Plan places emphasis on stakeholder capacity (see Environmental Foundation of Jamaica 2012, 2017). NGOs such as the Caribbean Coastal Area Management Foundation have a strong track record of implementing locally and internationally-funded community-based adaptation and capacity-building projects (see United Nations Development Programme 2017).

Capacity-building projects/programmes

Jamaica has participated in a number of regional projects and programmes for building capacity related to climate change, in which the Caribbean Community (CARICOM) and its Climate Change Centre (CCCCC) have been instrumental. One of three regional projects, the Caribbean Planning for Adaptation on Climate Change Project (CPACCP), ran from 1997 to 2004 and sought to 'build capacity in the Caribbean region for the adaptation to climate change impacts, particularly sea level rise' (Caribbean Community Climate Change Centre 2017b). Jamaica was one of 12 target countries, and, along with the Bahamas and Belize, was selected for the implementation of one of five pilot projects – climate change-related coral reef monitoring (ibid.). CPACCP was funded by the Global Environment Facility (GEF) and received US$5.6 million and was implemented by the World Bank, executed by the Organisation of American States and overseen by an ad-hoc committee of CARICOM (ibid.).

The Adaptation to Climate Change in the Caribbean (ACCC) regional project, which ran from 2001 till 2004, built on the activities and achievements of the CPACCP and filled gaps left by it. In addition to facilitating 'the transformation of the Regional Project Implementation Unit (RPIU) originally established through CPACC into a legal regional entity for climate change [i.e. the CCCCC]' and other activities, project funds were used to downscale climate scenarios for Jamaica (Caribbean Community Climate Change Centre 2017a). The ACCC Project was funded by the Canadian International Development Agency to the amount of CAD$3.5 million and was overseen by the World Bank and CARICOM (Caribbean Community Climate Change Centre 2017a, 2017b).

The Mainstreaming Adaptation to Climate (MACC) Project was the third regional initiative (2004–2007), aiming to 'mainstream climate change adaptation strategies into the sustainable development agendas of the small island and low-lying states of CARICOM'. The MACC Project 'adopted a learning-by-doing approach to capacity building, consolidating the achievements of CPACC and ACCC' (Caribbean Community Climate

Change Centre 2017c). Again, Jamaica was one of 12 participating countries. The project carried out vulnerability and capacity assessments of Jamaica's water sector, and developed a sectoral climate adaptation strategy (ibid.). The MACC Project was funded by GEF for US$5 million and was supported by in-kind contributions from the Government of the United States through its National Oceanic and Atmospheric Administration and from the Government of Canada. The project was implemented by the World Bank (ibid.).

Synthesizing the similarities and differences among country case studies

The brief exploration of the three countries – Bangladesh, Uganda and Jamaica – represents three different geographical configurations on three far-away continents. Bangladesh is a coastal country, bordered to the south by the Bay of Bengal, which is part of the Indian Ocean. Uganda is a landlocked country, with rich freshwater resources, particularly from Lake Victoria. Jamaica is a small island state in the Caribbean Bay.

In terms of development and level of income, the three countries also differ greatly, ranging from US$625 per capita income in Uganda to over $1500 in Bangladesh to about $5000 in Jamaica. The differences in geography and income level have implications for the impacts of and vulnerabilities to climate change. While Jamaica faces certain existential threats due to its physical location and because it is a small island developing state, the other two countries are LDCs and also suffer from socio-economic vulnerabilities, such as poverty and a lack of education.

In terms of economic vulnerabilities, Bangladesh is the most densely-populated and so will be exposed to far more loss and damage relative to the other two countries. Agriculture is the mainstay of Uganda's economy, while the ready-made garment sector is the dominant sector Bangladesh, and tourism is the main feature of the Jamaican economy.

However, each of these three countries is particularly vulnerable, with particular vulnerabilities to each of their agriculture and water sectors. In terms of climate disasters, however, Uganda suffers more from floods, droughts and seasonally variable heavy rainfalls. Bangladesh, on the other hand, suffers more from floods, cyclones and storm surges, as well as from erratic rainfalls. Jamaica, like Bangladesh, is extremely vulnerable to hurricanes, storm surges and depressions.

There are similarities and differences across the three countries in terms of policy-institutional frameworks. All three have developed elaborate policies, plans, visions and institutional mechanisms to cope with climate change. However, unlike in Bangladesh and Jamaica, Uganda has a national level climate change committee and advisory group that includes all stakeholders – even civil society representatives. Uganda is distinct in two other aspects – its development of a GHG Inventory Management

System, and climate budget mainstreaming down to the district level, with specific guidelines. Bangladesh has developed a Climate Fiscal Framework, though it has operationalized, and has a climate change trust that includes civil society and NGO representatives. Bangladesh also differs from Uganda and Jamaica in terms of its regular inclusion of non-governmental experts in the UNFCCC negotiations, and it has a vibrant, world-famous NGO sector in its development/environment sectors that is involved in climate change issues. Unlike national-level projects and NGOs/networks in Bangladesh and Uganda, Jamaica has strong regional networks and solidarity in the Caribbean because of its geographical proximity to and similar vulnerabilities as other Caribbean locations.

As for international support, Bangladesh and Uganda appear to depend more on bilateral sources of funding, such as DFID and DANIDA, while Jamaica gets more funding for its regional projects from the World Bank, CARICOM and the Inter-American Development Bank.

In terms of capacity building efforts, Bangladesh has a distinct advantage in the area of disaster management and rehabilitation areas as a country that has historically been extremely vulnerable to climate disasters. This adds an edge in its leadership of adaptation efforts among developing countries and LDCs. However, as is also the case in Uganda, DRM/DRR and climate change are dealt with by two different ministries. This stands in the way of achieving synergy and coordination in both countries. Both also suffer from lack of funds, expertise in technical areas and coordination among MADs. This may be due to their large size as compared to the compact, small size of Jamaica, which facilitates regional assessment of impacts and vulnerabilities, and capacity building.

References

Africa Climate Change Resilience Alliance (ACCRA). (2017). *Planning for NDC implementation in Uganda: MRV readiness assessment.* Available at; https://cdkn.org/project/africa-climate-change-resilience-alliance

Ahmed, A. U., Huq, S., Nasreen, M., & Hassan, A. R. (2015). Sectoral inputs towards the formulation of Seventh Five Year Plan 2016–2021. Climate change and disaster management. Available at: plancomm.gov.bd: www.plancomm.gov.bd/wp-content/uploads/2015/02/11a_Climate-Change-and-Disaster-Management.pdf

Ahmed, M., & Suphachalasai, S. (2014). *Assessing the costs of climate change and adaptation in South Asia.* Mandaluyong City, the Philippines: Asian Development Bank.

Amone, W. (2014). *Agricultural productivity and economic development in Uganda: An inclusive growth analysis.* Mbarara University of Science and Technology, Uganda.

Caribbean Community Climate Change Centre. (2017a). Adaptation to Climate Change in the Caribbean Project (ACCC). Available at: www.caribbeanclimate.bz/closed-projects/2001-2004-adaptation-to-climate-change-in-the-caribbean-project-accc.html (accessed 15 July 2017).

Caribbean Community Climate Change Centre. (2017b). Caribbean Planning for Adaptation to Climate Change Project (CPACC). www.caribbeanclimate.bz/closed-projects/1997-2001-caribbean-planning-for-adaptation-to-climate-change-project-cpacc.html (accessed 15 July 2017).

Caribbean Community Climate Change Centre. (2017c). Mainstreaming and Adaptation to Climate Change (MACC). www.caribbeanclimate.bz/closed-projects/2004-2007-mainstreaming-adaptation-to-climate-change-macc.html. (accessed 15 July 2017).

CARIBSAVE Partnership. (2012). *CARIBSAVE climate change risk profile for Jamaica*. Christ Church, Barbados: CARIBSAVE Partnership.

DOE (Department of Environment). (2007). *Climate change and Bangladesh*. Dhaka: Ministry of Environment and Forest, Government of Bangladesh.

Environmental Foundation of Jamaica. (2012). *EFJ strategic plan 2013–2015*. Kingston: Environmental Foundation of Jamaica.

Environmental Foundation of Jamaica. (2017). SCCAF projects awarded as at February 2017. Available at: www.efj.org.jm/sccaf-projects-awarded-february-2017 (accessed 31 July 2017).

FAO. (2006). Livelihood adaptation to climate variability and changes in drought-prone areas of Bangladesh. Rome: FAO. Available at: www.fao.org/3/a-a0820e.pdf

FAO. (2012). *FAOSTAT Database*. Rome: FAO. Available at: www.faostat.org/

FAO. (2015). *Cost benefit analysis (CBA) of climate change adaptation and prioritization in agriculture, environment and water sectors in Uganda. vulnerability assessment report*. Rome: FAO.

Future Climate for Africa. (2016). Africa's climate: Helping decision makers make sense of climate information. Uganda country fact sheet. Current and projected climate. Available at: www.futureclimateafrica.org/wp-content/uploads/2016/11/africas

GED. (2015). *Seventh Five- Year Plan FY2016–FY2020 (2015): Accelerating growth, empowering citizens*. Dhaka: General Economic Division, Planning Commission. Government of the People's Republic of Bangladesh.

GoB (Government of Bangladesh), & UNDP (2009). *Policy study on the probable impacts of climate change on poverty and economic growth and the options of coping with adverse effect of climate change in Bangladesh*. Dhaka: General Economic Division, Planning Commission. Available at: www.climatechange.gov.bd/sites/default/ files/ GED_policy_report.pdf

Gobeshona. (n.d.). Making research on climate change in Bangladesh more effective. Available at: http://gobeshona.net/#sthash.40Kwa1ol.dpbs (accessed 14 October 2017).

Government of Jamaica. (2000). *Jamaica's First National Communication to the United Nations Framework Convention on Climate Change (UNFCCC)*. Kingston: Ministry of Water and Housing & Meteorological Service of Jamaica.

Government of Jamaica. (2009). *Jamaica's national energy policy 2009–2030*. Kingston: Government of Jamaica.

Government of Jamaica. (2011). *The Second National Communication of Jamaica to the United Nations Framework Convention on Climate Change*. edited by Meteorological Service of Jamaica. Kingston: Meteorological Service of Jamaica.

Government of Jamaica. (2013). *Green Paper: Climate Change Policy Framework and Action Plan. edited by Land Ministry of Water, Environment and Climate Change*. Kingston: Government of Jamaica.

Government of Jamaica. (2016). *Biennial Update Report of Jamaica*. Bonn: United Nations Framework Convention on Climate Change Secretariat.

Government of Uganda. (2002). *Uganda First National Communication to UNFCCC*. Ministry of Water, Lands and Environment. Available at: http://unfccc.int/resource/docs/natc/uganc1.pdf

Government of Uganda. (2007). Climate change: Uganda national adaptation programmes of action. Available at: http://unfccc.int/resource/docs/napa/uga01.pdf

Government of Uganda. (2009). *The State of Uganda Population Report 2009, addressing the effects of climate change on migration patterns and women*. United Nations Population Fund. Available at: www. ibrary.health.go.ug/download/file/fid/1796

Government of Uganda. (2010). *Uganda vision 2040*. Kampala: National Planning Authority.

Government of Uganda. (2014a). *Second national communication*. Kampala: Ministry of Water and Environment, Climate Change Department.

Government of Uganda. (2014b). *The poverty status report 2014*. Kampala: Ministry of Finance, Planning and Economic Development (MFPED). Available at: http://planipolis.iiep.unesco.org/en/2014/poverty-status-report-2014-structural-change-and-poverty-reduction-uganda-6185

Government of Uganda. (2015a). *2015 annual statistical abstract*. Kampala: Uganda Bureau of Statistics (UBOS).

Government of Uganda. (2015b). *Uganda climate-smart agriculture country program 2015–2025*. Kampala: Ministry of Agriculture Animal Industry and Fisheries (MMAIF) and Ministry of Water and Environment (MWE).

Government of Uganda. (2015c). *Uganda's intended nationally determined contributions*. Kampala: Ministry of Water and Environment.

Government of Uganda. (2015d). *Second national development plan 2015/2016–2019/2020*. Kampala: National Planning Authority.

Government of Uganda. (2015e). *National climate change policy for Uganda*. Kampala: Ministry of Water and Environment.

Government of Uganda. (2016). *The 2014 national population and housing census main report*. Kampala: Uganda Bureau of Statistics (UBOS).

Government of Uganda. (2017). *Strategic Programme for Climate Resilience: Uganda Pilot Programme for Climate Resilience, April 2017*. Kampala.

Grey, O. (2017). Presentation: national climate change adaptation planning – Jamaica. Winnipeg: NAP Global Network.

History Cyclone Preparedness Programme (CPP)-Government of the People's Republic of Bangladesh (n.d.). Available at: www.cpp.gov.bd/site/page/7bd9cae0-55b4-41cd-a3fb-776243dcd068/History

Huq, S. (2001). Climate change and Bangladesh. *Science*, *294*(5547), 1617.

Huq S., & Khan M. R. (2017). Planning for adaptation in Bangladesh: Past, present and future. ICCCAD Policy Brief. Dhaka: ICCCAD.

Huq, S., & Rabbani, G. (2011). Climate change and Bangladesh: Policy and institutional development to reduce vulnerability. *Journal of Bangladesh Studies*, *13*, 1–10.

Hundsbæk, R. P., Spichiger, R., Alobo, S., & Kidoido M. (2012). Land tenure and economic activities in Uganda: A literature review. Working Paper 2012: 13. Danish Institute for International Studies (DIIS).

IPCC. (2007). *Climate change 2007: impacts, adaptation and vulnerability: contribution of Working Group II to the Fourth Assessment Report of the Intergovernmental Panel on Climate Change*. Cambridge: Cambridge University Press. Available at: www.ipcc.ch/pdf/assessment-report/ar4/wg2/ar4_wg2_full_report.pdf

IPCC. (2014). *Climate Change Synthesis Report. Fifth Assessment Report (AR5)*. United Nations Framework Convention on Climate Change. Cambridge: Cambridge University Press.

Jamaica Information Service. (2017). Vision 2030 Jamaica – National Development Plan. Kingston: Government of Jamaica. Available at: http://jis.gov.jm/features/vision-2030-jamaica-national-development-plan/ (accessed 29 July 2017).

Kreft, S., Eckstein, D., & Melchior, I. (2017). *Global Climate Risk Index*. Bonn: Germanwatch.

Lyew-Ayee Jr, P. (2015). Son of a beach: Jamaica and the coastal zone. 10th Annual EFJ Public Lecture. Montego Bay: Environmental Foundation of Jamaica.

Maplecroft, V. (2011). *Climate Change Vulnerability Index, 2015*. Climate Change and Environmental Risk Atlas. Available at: reliefweb.int/map/world/world-climate-change-vulnerability-index-2015

McCalla, W. (2012). *Review of policy, plans, legislation and regulations for climate resilience in Jamaica*. Kingston: Planning Institute of Jamaica.

McSweeny, C., New, M., & Lizcano, G. (2010). UNDP climate change country profiles: Uganda. Available at: www.geog.ox.ac.uk/research/climate/projects/undp-cp/index.html?country=Uganda&d1=Observed&d2=Extremes&d3=Time series

Meteorological Service of Jamaica. (2002a). About the Meteorological Service of Jamaica. Government of Jamaica. Available at: www.metservice.gov.jm/aboutus.asp (accessed 20 July 2017).

Meteorological Service of Jamaica. (2002b). Climate of Jamaica. Government of Jamaica. Available at: www.metservice.gov.jm/climate.asp (accessed 20 July 2017).

Meteorological Service of Jamaica. (2002c). Contact us. Government of Jamaica. Available at: www.metservice.gov.jm/contactus.asp (accessed 20 July 2017).

Meteorological Service of Jamaica. (n.d.). Jamaica climate. Government of Jamaica. Available at: http://jamaicaclimate.net/ (accessed 31 July 2017).

Ministry of Economic Growth and Job Creation. (2017). Divisions and branches. Government of Jamaica. Available at: www.mwh.gov.jm/#!/divisions (accessed 17 July 2017).

Ministry of Local Government and Community Development. (2013). About us. Government of Jamaica. Available at: www.localgovjamaica.gov.jm/aboutus.aspx. (accessed 30 July 2017).

Ministry of Science Energy and Technology. (2017). National energy policy. Government of Jamaica. Available at: http://mset.gov.jm/national-energy-policy. (accessed 22 July 2017).

Ministry of Water and Environment. (2014). *Guidelines for the integration of climate change in sector plans and budgets*. Kampala: MoWE.

Ministry of Water and Environment. (2015). *Economic assessment of the impacts of climate change in Uganda. Final Report. November 2015. Study conducted with support from CDKN and DFID Uganda*. Kampala; Government of Uganda.

Ministry of Water and Environment. (2017). *Joint Water and Environment Sector Support Programme 2013–2018: Consolidation Strategy and Plan. Final Draft*. Kampala; Government of Uganda.

MoDMR (Ministry of Disaster Management and Relief.) (2016). *National plan for :disaster management (2016–2020): Building resilience for sustainable human development*. Dhaka, Bangladesh.

MoEF. (2002). Initial National Communication of Bangladesh to the United Nations Framework Convention on Climate Change. Government of the People's Republic of Bangladesh. Available at: http://unfccc.int/resource/docs/natc/bgdnc1.pdf

MoEF. (2005). *National Adaptation Programme of Action*. Dhaka: Government of the People's Republic of Bangladesh.

MoEF. (2007). *National capacity self-assessment: Capacity development action plan for sustainable environmental governance*. Dhaka: Government of the People's Republic of Bangladesh.

MoEF. (2009, September). Bangladesh Climate Change Strategy and Action Plan 2009. Available at: climatechangecell.org: www.climatechangecell.org.bd/Documents/climate_change_strategy2009.pdf

MoEF. (2012). Second National Communication of Bangladesh to the United Nations Framework Convention on Climate Change. Available at: unfccc.int: http://unfccc.int/resource/docs/natc/bgdnc2.pdf

MoEF. (2014). *Climate fiscal framework*. Finance Division, Ministry of Finance. Dhaka: Government of the People's Republic of Bangladesh.

MoEF (2017). *The Country Investment Plan (CIP) for Environment, Forestry and Climate Change (EFCC)*. Dhaka: GOB.

Nachmany, M., Fankhauser, S., Davidová, J., Kingsmill, N., Landesman, T., ... & Townshend, T. (2015). Climate change legislation in Jamaica. In *2015 Global Climate Legislation Study: A Review of Climate Change Legislation in 99 Countries*. London: London School of Economics.

National Environment Management Authority (NEMA). (2010). *Uganda state of the environment report 2010*. Kampala: Government of Uganda.

Notre Dame Global Adaptation Initiative. Notre Dame Global Adaptation Index: Uganda. Available at: http://index.gain.org/country/uganda. (accessed 14 October 2017).

OECD. (2012). OECD Statistics. Available at: https://stats.oecd.org/

Office of Disaster Preparedness and Emergency Management. (2008a). How ODPEM actively prepares Jamaica for disasters. Government of Jamaica. Available at: http://odpem.org.jm/BePrepared/HowODPEMPreparesJa/tabid/72/Default.aspx (accessed 31 July 2017).

Office of Disaster Preparedness and Emergency Management. (2008b). National disaster plans. Government of Jamaica. Available at: http://odpem.org.jm/DisastersDoHappen/NationalDisasterPlans/tabid/57/Default.aspx (accessed 31 July 2017).

Office of Disaster Preparedness and Emergency Management. (2008c). Organizational structure. Government of Jamaica. Available at: http://odpem.org.jm/AboutUs/OrganizationStructure/tabid/90/Default.aspx (accessed 31 July 2017).

Office of Disaster Preparedness and Emergency Management. (2008d). Types of hazards and disasters. Government of Jamaica. Available at: http://odpem.org.jm/AboutUs/OrganizationStructure/tabid/90/Default.aspx. (accessed 31 July 2017).

Office of Disaster Preparedness and Emergency Management. (2008e). Who we are. Government of Jamaica. Available at: http://odpem.org.jm/AboutUs/WhoWeAre/tabid/87/Default.aspx (accessed 31 July 2017).

Office of the Prime Minister. (2017). Ministry of Economic Growth and Job Creation. Government of Jamaica. Available at: http://opm.gov.jm/portfolios/ministry-of-economic-growth-and-job-creation/ (accessed 17 July 2017).

PANOS Caribbean. (2016a). #1point5toStayAlive: The Caribbean's Climate Justice Hub. Available at: http://panoscaribbean.org/mission (accessed 31 July 2017).

PANOS Caribbean. (2016b). Mission + vision. Available at: http://panoscaribbean.org/mission (accessed 31 July 2017).

Planning Institute of Jamaica. (2008). Main functions and responsibilities. Government of Jamaica. Available at: www.pioj.gov.jm/AboutUs/Functions/tabid/108/Default.aspx (accessed 17 July 2017).

Planning Institute of Jamaica. (2009). Press release: Vision 2030 Jamaica plan ready for implementation. Available at: www.vision2030.gov.jm/News-and-Events/Press-Releases/xmmid/470/xmid/436/xmview/2 (accessed 19 July 2017).

Planning Institute of Jamaica. (2012). Welcome to Vision 2030 Jamaica. Government of Jamaica. Available at: www.vision2030.gov.jm/ (accessed 17 July 2017).

Planning Institute of Jamaica. (2015). Medium Term Socio-Economic Policy Framework (MTF). Government of Jamaica. Available at: www.vision2030.gov.jm/Medium-Term-Socio-Economic-Policy-Framework. (accessed 30 July 2017).

Planning Institute of Jamaica. (2017). Vision 2030 Jamaica – National Development Plan: dashboard of indicators. Available at; http://devinfolive.info/dashboard/Jamaica_vision2030/index.php (accessed 19 July 2017).

Planning Institute of Jamaica. (n.d.). Overview: Vision 2030 Jamaica National Development Plan. In *Country planning cycle database*. Geneva: World Health Organization.

Richards, A. (2008). *Development trends in Jamaica's coastal areas and the implications for climate change*. Kingston: Planning Institute of Jamaica.

Rozenberg, J., & Hallegatte, S. (2015). *Shock waves: Managing the impacts of climate change on poverty*. Washington, DC: World Bank.

Statistical Institute of Jamaica. (2017). Population of Jamaica. Available at: http://statinja.gov.jm/demo_socialstats/newpopulation.aspx (accessed 20 July 2017).

Third National Communication to the UNFCCC. (n.d.). Available at: www.bd.undp.org/content/bangladesh/en/home/operations/projects/environment_and_energy/third-national-communication-to-the-unfccc.html

Twinomuhangi, R., Buliung, R. N., Nakitto, M. T., & Lett, R. (2014). Application of geographic information systems methodology to injury surveillance in Uganda. *International Research Journal of Medicine and Medical Sciences*, 2(2), 20–39.

Twinomuhangi, R., & Monkhouse C. (2015). Economic assessment of the impacts of climate change in Uganda: Key results. CDKN. Available at: https://cdkn.org/wp-content/uploads/2015/11/UGANDA_Economic-assessment-of-climate-change_WEB.pdf

Uganda Bureau of Statistics. (2014). Statistical abstracts 2014. Kampala: Government of Uganda.

UNDP (2013). *Climate risk management for sustainable crop production in Uganda: Rakai and Kapchorwa Districts*. New York. United Nations Development Programme (UNDP), Bureau for Crisis Prevention and Recovery (BCPR). Available at: www.iisd.org/pdf/2013/crm_uganda.pdf

UNDP. (2015). *Uganda sustainable development report 2015. strategies for sustainable landuse and management. final report*. Ministry of Finance, Planning and Economic Development. Kampala: Government of Uganda.

UNDP. (2016). Uganda launches digital Greenhouse Gas Inventory System. Available at: www.ug.undp.org/content/uganda/en/home/presscenter/pressreleases/2016/10/21/uganda-launches-digital-greenhouse-gas-inventory-system-.html

UNDP. (2017). CBA Jamaica: Increasing community adaptation and ecosystem resilience to climate change in Portland Bight (CCAM). Available at: www.adaptation-undp.org/projects/spa-cba-jamaica-increasing-community-adaptation-and-ecosystem-resilience-climate-change (accessed 16 July 2017).

UNEP. (n.d.). Resilience: Bangladesh leads way with national adaptation plan NAP process. Available at: www.unep.org/climatechange/resilience-bangladesh-leads-way-national-adaptation-plan-nap-process

United Nations Framework Convention on Climate Change. (2016, 27 April). Status of ratification of the Kyoto Protocol. Available at: http://unfccc.int/kyoto_protocol/status_of_ratification/items/2613.php

United States Census Bureau. (2017). State population totals tables: 2010–2016. United States Census Bureau. Available at: www.census.gov/data/tables/2016/demo/popest/state-total.html

USAID. (2011). *Climate change and conflict in Uganda: The cattle corridor and Karamoja*. African and Latin American Resilience to Climate Change Project (ARCC). Discussion Paper for Office of Conflict Management and Mitigation.

USAID. (2013): Uganda climate change vulnerability assessment report, prepared by Caffrey, P., Finan, T., Trzaska, S., Miller, D., Laker-Ojok, R. and Huston, S. Washington, DC: USAID.

USAID. (2015). *Climate change information fact sheet: UGANDA*. Washington, DC: USAID.

Williams-Raynor, P. (2015a). Government looks to merge climate change division with Met Office. *The Jamaica Gleaner*. Available at: http://jamaica-gleaner.com/article/news/20151202/government-looks-merge-climate-change-division-met-office (accessed 30 July 2017).

Williams-Raynor, P. (2015b). Jamaica gets going on emissions reduction. *The Jamaica Gleaner*, 2 April. Available at: www.jamaica-gleaner.com/article/news/20150402/jamaica-gets-going-emissions-reduction (accessed 3 April 2015).

Williams-Raynor, P. (2015c). Jamaica's climate policy gets cabinet approval. *The Jamaica Gleaner*, 11 March. Available at: http://jamaica-gleaner.com/article/news/20150311/jamaica%E2%80%99s-climate-policy-gets-cabinet-approval (accessed 15 March 2015).

Williams-Raynor, P. (2015d). New climate change publication caters to the blind. *The Jamaica Gleaner*, Available at: http://jamaica-gleaner.com/article/news/2015 0410/new-climate-change-publication-caters-blind (accessed 21 July 2017).

World Bank. (2012a). *World development indicators*. Washington, DC: The World Bank. Available at: http://bit.ly/1aS5CmL

World Bank. (2012b). *Uganda Country Environmental Analysis (CEA)*. Washington, DC: World Bank.

World Bank. (2015a). *Uganda climate profile*. Washington, DC: World Bank.

World Bank. (2015b). *Uganda strategic climate diagnostic*. Washington, DC: World Bank.

World Bank. (2016). World GDP per capita ranking 2016. Available at: http://uganda.opendataforafrica.org/sijweyg/world-gdp-per-capita-ranking-2016-data-and-charts-forecast

World Travel & Tourism Council. (2013). *Benchmarking travel & tourism in Jamaica: How does travel & tourism compare to other sectors*? London: World Travel & Tourism Council.

6 Lessons learned from agency initiatives on capacity building

With Julianna Bradley

Introduction

For decades, foreign assistance and cooperation agencies from around the world have been focused on building capacity in developing countries. What are they funding? How is it changing? What do we know about whether it's working? For this chapter, we undertook a comparative review of approaches to capacity building in the area of climate change that have been taken by development cooperation agencies around the world. Our review included national 'bilateral' agencies like UK Aid (DFID), USAID, Canadian CIDA, Swedish SIDA and Norway's NORAD, and 'multilateral' regional banks and funds administered by the World Bank, where there has been a lot of experience in this area. We also looked at the Global Environment Facility and its funding entities, like the UNDP, the UNEP, the Adaptation Fund and the Least Developed Countries Fund. Our analysis looked at projects from 2006 to the time of our searches, which were in the first half of 2017.

We searched agencies' project and activity databases for those that addressed climate and capacity building, seeking to understand the processes and tools they brought to bear. In doing so, we sought to understand six questions. (1) Where is the most money going?; (2) Who are the most frequent implementers?; (3) Who are the most frequent recipients?; (4) Which of the 15 capacity-building priority areas set out by the UNFCCC (see Chapter 2) seem to be getting the most attention by these agencies? That is, which have the most projects and the most dedicated funding?; (5) Who is reporting well on capacity-building efforts? What kinds of information are available, and what's missing?; (6) Finally, can we identify some best practices in this area, including the actual investments being made and process, and how efforts are being monitored and reported?

The chapter begins with a review of capacity-building initiatives under bilateral agencies, with cases of DFID, CIDA, NORAD/SIDA and USAID, the aid agencies from the UK, Canada, Norway, Sweden and the USA, respectively. For each, we built a database of projects including the keywords 'climate' and 'capacity building' over the period 2006–2017. We summed the projects and where possible describe major trends in each agency's efforts. Next, we review capacity building initiatives under multilateral agencies: the Asian Development Bank, the African Development Bank, the Inter-American Development Bank, the World Bank and the Pilot Project for Climate Resilience they administer. Finally, we review the capacity-building initiatives under the Global Environment Facility, the Least Developed Countries Fund and the Adaptation Fund. Much more work in this area has taken place through UNDP and the UNEP, which implement some of the other funds, so we look into their capacity-building efforts.

For all the agency summaries, we attempted to roughly follow a similar format. First, we introduce the agency and its stated goals, including on climate change. Next, we discuss how we retrieved their project information, with details on the databases we searched, keywords we used, and when we conducted the search. Then we describe how many projects we uncovered and examined, their size, and we list the top 2–5 projects, to give a flavour for what they're funding. We describe that contributor's main recipient nations, and which implementing agencies tend to carry out their projects. We describe the categories of projects they conduct. We review the stated goals we could uncover of each agency's climate change-related capacity-building initiatives. And we review the means by which they sought to achieve those goals. We end with reflections on the questions above that drive this chapter: Which capacity-building priority areas are getting the most attention and who is reporting well on capacity-building efforts? Concluding our analysis, we also discuss which issues tend to get addressed in this work, and where, and what gets left out.

Capacity-building initiatives under bilateral agencies

UK Aid/DFID: Department for International Development (UK)

The DFID is the UK's lead aid disbursing agency, responsible for 80 per cent of the country's Official Development Assistance (ODA) in 2015 (National Audit Office 2017). The DFID's stated goal is to 'end extreme poverty'. With this aim, DFID works directly in developing countries as well as giving aid to multi-country global programmes and to multilateral development agencies. Last, but prominent, in its list of seven top responsibilities is 'helping to prevent climate change and encouraging adaptation and low-carbon growth in developing countries' (DFID 2017c) We attempted to characterize the climate change capacity-building assistance among the UK's commitment of spending 0.7 per cent of national income for ODA.

To do so, we went to the Development Tracker, the DFID's online tool for exploring UK aid projects. Keywords 'climate capacity building' between 2006 and 2016 returned 50 projects, a later search after their website changed returned 236 results. This shows the difficulty in producing a definitive universe of projects from changing websites, a problem with all the agencies, and a reminder that this chapter's analysis should be seen as indicative, not definitive. Of the 50 projects analysed, 4 were excluded as irrelevant for climate change, and of the rest, only 6 projects (13 per cent) provided evaluation reports (2 reported by DFID and 4 reported by the UK Department of Energy and Climate Change). Still, some observations can be made about the projects we were able to identify. First, the vast majority of climate capacity building projects were under £1 million pounds (36 projects), with 24 of those being for less than £100,000. In fact, eight projects were tiny, only committing under £20,000 for their efforts. Of the ten projects listed as over £1 million, only three were over £10 million, and just one was over £100 million (Table 6.1).

Importantly, the most frequent recipients of capacity-building project funds were *British* civil society organizations, especially of the smaller budgetary awards: Oxfam Great Britain (15 projects), CAFOD, the Catholic Agency for Overseas Development (11 projects), and the Foreign & Commonwealth Office (10 projects). The most frequent *implementers* of the projects were local civil society organizations in developing countries, often related to religious and cultural organizations. The projects were distributed across the world, in Asia, Africa and Latin America.

Table 6.1 Sample major capacity-building projects, UK DFID

Forest Governance, Markets and Climate (2011–2021)

- Finance: $220,902,874 (£162,767,284)
- Recipient country: Multiple recipients
- Funding agency: DFID
- Implementing Agency: Centre for International Forestry Research (CIFOR) (GBP), Chatham House Enterprises, European Forest Institute, European Timber Trade Federation, FERN, IIED, Well Grounded Ltd
- Executing Agency: DFID

(DFID 2017)

International Carbon Capture and Storage Capacity Building (2012–2015)

- Finance: $81,430,200.00 (£60,000,000)
- Recipient country: South Africa
- Funding agency: International Climate Fund
- Implementing Agency: Not Stated
- Executing Agency: DFID

(DECC 2016)

Lower Limpopo River Valley Flood Reconstruction and Climate Resilience Project (2014–2017)

- Finance: $19,020,737.55 (£14,015,000)
- Recipient country: Mozambique
- Funding agency: DFID
- Implementing Agency: IBRD (HSBC) C/O The World Bank
- Executing Agency: DFID

(DFID 2017)

The objectives of DFID's capacity-building projects were based on advocacy and intersectional issues such as women's rights. Several projects focused specifically on capacity building in the communication sector. To achieve these objectives the DFID conducted training programmes, implemented education initiatives and supported advocacy and research projects.

The Canadian International Development Agency: CIDA

The Canadian International Development Agency (CIDA) was the federal Canadian administrative body for foreign aid programmes in developing countries. In 2013, CIDA was folded into operations of the Department of Foreign Affairs, later renamed Global Affairs Canada. At the centre of CIDA's objectives is the elimination of barriers to equality and creating better opportunities for women and girls as agents of change. CIDA frequently argues that women as key recipients of their aid have immense potential to improve their families, communities and countries (CIDA 2017). With regards to climate change, CIDA centres its approach around issues of gender equality. Women and girls living in poverty are at greater risk of climate-related challenges, as they are often primary food producers and providers of water and cooking fuel. These resources are becoming more unpredictable and scarcer due to extreme weather and changes in water availability. Further, CIDA argues the inclusion of women in active roles in designing and developing strategic responses to climate change is essential to address the stresses of a changing climate for all those living in poverty (ibid.). With this commitment in mind, we considered the nature of CIDA's funding programmes specific to capacity building.

Global Affairs Canada's Report on Plans and Priorities 2016–17 prioritizes Canada's leadership role in the international arena, intending to 'enhance Canada's leadership on clean energy and international efforts to combat climate change and mitigate its impact on the most vulnerable, including through climate finance mechanisms' (Global Affairs Canada 2016–2017). This funding aims to aid such communities in mitigating and adapting to climate change and make a transition to a low-carbon, climate-resilient economic system. We searched Canada's International Development Project Browser for keywords 'capacity building' and narrowed the sectors to General Environmental Protection and the timeline to projects initiated between 2007 and 2017. This search produced a total of 17 projects with a total of CN$3,041,552,100 (US$2,471,500,604). Four projects had funding totals below CN$1 million. The largest projects including capacity building and climate change are shown in Table 6.2.

A significant project is the World Bank BioCarbon Plus Fund, which provides financing for projects that sequester or conserve greenhouse gases. One project received CN$4.5 million (US$3,658,950.81) from CIDA for capacity building in agricultural and forestry sectors by creating the Sustainable Agricultural Land Management carbon accounting methodology

Table 6.2 Sample major capacity-building projects, CIDA

Energy Sector Capacity Building (2013–2016)

- Finance: $12,694,500.01 (CN$15,500,000.01)
- Recipient country: Tanzania
- Funding agency: IBRD Trust Funds – World Bank
- Implementing Agency: World Bank
- Executing Agency: Ministry of Energy and Minerals, the Tanzania Electric Supply Company Limited, the Tanzania Petroleum Development Corporation, and the Energy and Water Utilities Regulatory Authority

(Global Affairs Canada 2017)

World Bank BioCarbon Plus Fund (2011–2020)

- Finance: $3,685,500.00 (CN$4,500,000)
- Recipient country: Multiregional
- Funding agency: IBRD Trust Funds – World Bank
- Implementing Agency: World Bank Institute
- Executing Agency: Word Bank

(Global Affairs Canada 2017)

to allow smallholder farmers to benefit from payments from accessing the carbon market.

Environmental sustainability and climate change adaptation are targeted through issues of equitable participation in livelihood production using natural resources and increasing the ability of women living in poverty to build capacity to climate change. CIDA sought to meet these markers of success through the implementation of gender equity-building training, programmes and partnerships that increased the participation of women in knowledge production, skills building and leadership. CIDA worked with local partners, often government organizations, NGOs, and existing programmes in order to increase the ability for increased livelihood production through the means of natural resource management. CIDA considers that the participation of women in livelihood production and natural resource management can enhance climate change adaptation.

United States Agency for International Development: USAID

Because data on US government websites became opaque on climate initiatives in 2017 after the inauguration of Donald Trump, historical USAID projects were scraped from the OECD-Creditor Reporting System site, using the 'Rio Marker' for climate change adaptation and mitigation, 'principal objective' (RM=2), and a search for the term 'capacity building'. For 2015 alone, 139 projects were returned. Most of these projects were Peace Corps projects:

> New Small Project Assistance program award with Peace Corps. The SPA Program enables USAID to have direct development investments

> and impact in selected priority areas, while supporting U.S. Peace Corps volunteers to engage community members in participatory processes that contribute to capacity building and sustainable development.

Also, some global/regional services provided to USAID implementers, such as providing 'on-demand environmental compliance, management, capacity-building and sound design support to USAID's Environmental Officers, to USAID Missions and other operating units, and to their projects and programs'. Because they were gathered in a different way than other agencies' projects, and because of known projects with the Adaptation Marker system (AdaptationWatch 2016; Roberts & Weikmans 2017), these findings seem to be of a different nature.

One relevant initiative was the Climate Change Resilient Development (CCRD) effort, a 'global, four-year project in support of USAID's Global Climate Change Office'. CCRD's climate adaptation-focused programmes operate in Asia, Africa, Eastern Europe, Latin America and the Caribbean. Employing a strategic development first framework, CCRD delivers guidance, technical assistance, and capacity building to USAID Missions and Bureaus, national governments, and local communities around the world to integrate climate change concerns into development policy, planning, and implementation. CCRD also explores emerging issues in climate adaptation by developing innovative programmes, such as the Climate Services Partnership (CSP), the High Mountains Adaptation Partnership (HiMAP), the Climate Resilient Infrastructure Services program (CRIS), and support for the preparation of National Adaptation Plans (NAPs) (USAID 2014) (Table 6.3).

CCRD was implemented by 'Engility/IRG and a consortium of private contractors, NGOs, and universities'. Likewise, The Climate Change Adaptation Assessments, Thought Leadership, and Learning (ATLAS) project works to improve the quality and effectiveness of USAID's development programmes aimed at reducing climate change risks through "tested and harmonized approaches to climate risk/adaptation-related assessments;" thought leadership on climate-resilient development; and capacity building of USAID and its implementing partners."

These two projects suggest that for USAID, capacity-building work mostly is for USAID staff, but with implications for initiatives around the world. These projects were all threatened at the time of this writing by uncertainty over funding, and also an ideological attack on the climate work promoted under the Obama Administration.

The Capacity Building of Cambodia's Local Organizations project was likewise set up to support USAID work on sustainable development through the use of local partnerships. The project was set up to support the development of new partnerships with local organizations and to strengthen the skills of existing partners across all technical sectors. The

Table 6.3 Sample major capacity-building projects, USAID

Climate Change Resilient Development (CCRD) – Clean Productive Environment Peru (2015–on-going)
- Finance: US$300,000
- Recipient country: Peru
- Funding agency: USAID
- Implementing Agency: Ministry of the Environment
- Executing Agency: USAID

(USAID 2014)

Capacity Building of Cambodia's Local Organizations Programme
- Finance: US$2,332,426.96
- Recipient country: Cambodia
- Funding agency: USAID
- Implementing Agency: the International Executive Service Corps (IESC) in partnership with Kanava International, LLC
- Executing Agency: Volunteers for Economic Growth Alliance (VEGA)

(VEGA 2015)

project was designed to help local organizations improve their financial management and human resource systems, develop strategic and operational plans, and strengthen their monitoring and evaluation systems.

These are classic capacity-building efforts, as seen in this book's earlier chapters.

Norway and Sweden: NORAD and SIDA

In per-capita terms, the Scandinavian countries are the most generous in the world for climate finance. In climate change adaptation finance, the Norwegian Agency for Development Cooperation (NORAD) and the Swedish International Development Cooperation Agency (SIDA) commit to providing funding for projects that build capacity while fostering sustainable development. NORAD operates under Norway's Ministry of Foreign Affairs and the Norwegian Ministry of Climate and Environment. NORAD recognizes the need for an increased focus and attention to adaptation to the challenges of climate change. It promotes development activities that help reduce vulnerability to climate change impacts, in a sustainable manner. Further, NORAD works primarily with partners in institutes of higher education, promoting the inclusion of institutional research and university programmes in development projects, as we propound in this book ('About Norad', n.d.; 'Adaptation to climate', n.d).

SIDA is committed to a central objective of reducing global poverty and a fundamental principle is the consideration of each recipient country's responsibility to adopt its own strategy for economic development and to combat poverty. Swedish development aid follows three thematic objectives: democracy and human rights, environment and climate change, and gender equality. Their focus in the field of environment and climate change involves the development of sustainable sanitation, fair and sustainable use of natural resources, and improving policy and legal frameworks. SIDA works directly with recipient countries' development aid agencies, international bodies such as the UN and the World Bank, and several Swedish government agencies.

For NORAD, we used the keywords 'capacity building', which produced 35 results, 13 of which have relevance to climate change and were initiated after 2006 (Table 6.4). All of the projects in the data set had partnerships with universities. Total funding for these projects was 222.5 million Norwegian Krones, or roughly US$29 million. For SIDA, we used the key words 'capacity building climate change' and received seven results. The total funding for these projects was 21,899,017 Swedish Kronas (roughly US$2.8 million).

All of the implementing agencies of the NORAD-funded projects are institutes of higher education, both in Norway and in the recipient countries. The Norwegian University of Life Sciences is an implementing agency for all but one of the projects in the dataset. Typically, each recipient country includes at least one university as an implementing agency, and in some cases several are included. The implementing agencies for SIDA-funded projects

Table 6.4 Sample major capacity-building projects, NORAD and SIDA

Regional Capacity Building for Sustainable Natural Resource Management and Agricultural Productivity under Changing Climate (2013–2018)

- Finance: $2,297,354.58 (18,000,000 NOK)
- Recipient country: Several countries in the East African Region
- Funding agency: NORAD
- Implementing Agency: NORAD
- Executing Agency: Makerere University, Uganda; University of Juba, South Sudan; Addis Ababa University, Ethiopia; Norwegian University of Life Sciences, Norway

(NORAD 2013a)

Incorporating Climate Change into Ecosystem Approaches to Fisheries and Aquaculture Management in Sri Lanka and Vietnam (2013–2018)

- Finance: $2,297,354.58 (18,000,000 NOK)
- Recipient country: Sri Lanka, Vietnam
- Funding agency: NORAD
- Implementing Agency: NORAD
- Executing Agency: Nha Trang University, Vietnam; University of Ruhuna, Sri Lanka; University of Tromsø, Norway; University of Bergen, Norway

(NORAD 2013b)

Indaba Agricultural Research Institute (2013–2017)

- Finance: $1,021,288.31 (8,137,321 SEK)
- Recipient country: Zambia
- Funding agency: SIDA
- Implementing Agency: Unclear
- Executing Agency: SIDA

(OpenAid n.d.)

are less clear. One project lists a distinct group, the World Bank, but every other project in the dataset only includes vague descriptions such an 'international NGO' or 'university, college, or other research institute'. The project *European Capacity Building Initiative* (with whom the authors of this book have worked) documentation lists Oxford Climate Policy as a 'cooperation partner' (NORAD 2013a). Other projects either do not provide documentation or do not list partnerships within the documentation.

As all of NORAD's projects are implemented with partnerships with universities, the objectives do seem to correlate with educational and institutional capacity building. Several projects aim to strengthen the human and institutional capacities of partner institutions in specific climate change-related fields such as water management, agriculture and environmental policy. One project, *Sustainable Natural Resource Management for Climate Change Adaptation in the Himalayan Region*, aims to 'build human capacity in order to impact the formulation and implementation of policies focused on sustainable resource management and climate change adaptation', through enhancing higher education and the research capacity related to climate change in a coordinated network between partner universities (NORAD 2013c). A general theme among all of the projects is the promotion of resilience in climate change adaptation in recipient countries through the strengthening of higher education in relevant fields. The specific means included strengthening the capacity to teach, research and supervise students on climate-specific issues, such as water management and sustainable development. Further, exchange between institutions among partner universities was initiated through the creation of joint databases and shared research projects. Several projects included the initiation of new PhD and Master's programmes in the field of climate change and development, or strengthening such programmes where they already existed. These programmes in higher education are intended to work in collaboration with actors outside of the university system, such as vulnerable communities and people working in natural resource management.

SIDA-funded projects also mentioned the intention of institutional capacity building in higher education, but they also discuss climate communications and diplomacy as well. The project *European Capacity Building Initiative* intends to build and sustain the capacity of communication between developing and developed countries in climate change negotiations ('ECBI phase 4', 2014). This initiative focuses on the capacity of negotiators from Least Developed Countries (LDCs) to originate ideas and to make impactful submissions and contributions to the talks (Benito Mueller, Director, pers. comm.). In several SIDA projects, the central objective is to build the capacity of the media to access and cover environmental issues and the challenges of climate change. Finally, similarly to NORAD, SIDA aims to enhance the linkages between resource groups through regional science-technology-policy dialogues. SIDA did not describe the means to carry out these implementations clearly. Often project descriptions did not mention intentions for implementation or were very vague, citing general intention such as 'strengthen capacity' (which was not unique to this agency). One project, *Building Long-Term Capacity for Managing and Adapting to Climate Change in Asia and Africa*, included a six-part programme with the intention to create resource groups, research fellowships and workshops to develop the research capacity of resource groups at the national and regional levels to analyse and assess the risk associated with global climate change (Chiang 2008). However, no other project included similar descriptions.

Overall, these bilateral agencies seem to have incorporated some climate change capacity-building efforts into their portfolios, but their approaches appear to vary substantially. For example, the UK projects were largely conducted through civil society organizations both at home and in the project's host country, while the Norwegian ones worked through universities, at both ends. It can be said that these agencies have fighting poverty and economic development as their core missions, and in historical terms climate change has only recently appeared in their planning. However, all five agencies do express in their public documents that helping developing nations prepare for climate risks is crucial to their missions of economic development and poverty reduction. Each sees a slightly different approach to advancing that goal. In terms of transparency, there were some serious deficits of information provided, and in some cases difficulty in ascertaining aid project information relevant to climate. Agencies provided some information relevant to each project, but they rarely included in-depth documentation regarding allocation of funds and project implementation.

Capacity-building initiatives under multilateral agencies

Asian Development Bank (ADB)

The Asian Development Bank's (ADB) stated mission is to free the Asia-Pacific region from poverty. Of the ADB's 67 member countries, 48 are from within Asia and the Pacific, and 46 are designated as developing country members (DMCs). As a multilateral development finance institution, the ADB dispenses loans, technical assistance and grants to member governments who are also shareholders of the bank. The ADB also provides equity investments and loans to private enterprises in developing country members. Its total lending in 2016 amounted to $31.70 billion.

The ADB pursues three development-oriented agendas of inclusive economic growth, environmentally sustainable growth, and regional integration. Its stated core sectors of focus under these agendas are infrastructure, environment, regional cooperation and integration, finance sector development, and education. The ADB's environmental agenda is contingent on and inextricable from its driving purpose of economic development (Asian Development Bank 2011). Its central objective is as follows:

> Environmental sustainability is a prerequisite for economic growth and poverty reduction in Asia and the Pacific ... environmental damage has started to threaten prospects for continued economic growth and poverty reduction. Maintaining natural capital must therefore be a crucial goal for countries.
>
> (ADB 2017c)

In order to examine the ADB's funding for capacity building, we explored the project database on their website under the 'Projects and Tenders' tab. Three searches conducted in 2017 over the period 1999–2016 for the words 'capacity building climate' yielded 56, 60 and 352 results. Two projects, 'Enhancing Gender and Development Capacity in DMCs' and 'Targeted Capacity Building for Mainstreaming Indigenous Peoples' Concerns in Development' did not explicitly have to do with climate change but were attributed to the ADB's Climate Change and Sustainable Development arm (ADB 2017a; ADB 2017b). Most of the capacity-building and climate-related projects had 'governance and capacity development' listed under the 'Drivers of change' filter category and 'environmentally sustainable growth' under 'strategic agendas'. However, we did not use these ADB-provided terms to filter our search.

We looked more closely at the 15 capacity-building and climate-related projects funded by the ADB with over $3 million of funding (Table 6.5). The largest were focused on building climate-resilient infrastructure in member countries. Of the climate-related capacity-building projects we found, 19 were in the agriculture, natural resources, and rural development sector, 17 were listed in the energy sector, 15 in public sector management, 14 in transport, and 16 in water and other urban infrastructure and services. Approximately a quarter of the capacity-building projects were regional. The remaining project recipients were distributed fairly evenly between South Asia, Southeast Asia, and China, with a handful going to island states in the Pacific. Country-specific projects all list one of the developing member countries' national ministries, offices, or committees as the implementing agency. Regional projects list the ADB office in the Philippines as the executing agency.

Although the ADB database describes many of the projects' intended outputs and rationales, the specific means of implementation are often left out. Coherent with the ADB's environmentally sustainable growth agenda, a theme running throughout many of the capacity-building projects is to mainstream and integrate climate resilience into a country's development and economic planning. Providing ADB developing member country governments and vulnerable communities with access to resources, 'tools and guidelines' is another means of implementation listed in several capacity-building projects, as well as setting up adaptation plans, communication networks, and monitoring, reporting, and evaluation (MRV) systems. The Promoting Urban Climate Change Resilience in Selected Asian Cities

Table 6.5 Sample major capacity-building projects, Asian Development Bank

Coastal Climate-Resilient Infrastructure Project (2011–on-going)

- Finance: $118.8 million
- Recipient country: Bangladesh
- Funding agency: Strategic Climate Fund, IFAD Grants, KfW Bankengruppe, Asian Development Fund, International Fund for Agricultural Development à Bangladesh
- Implementing Agency: ADB
- Executing Agency: Local Government Engineering Departments

(Asian Development Bank 2017a)

Rural Roads Improvement Project (2011–2016)

- Finance: $59.75 million
- Recipient country: Cambodia
- Funding agency: Asian Development Fund, Nordic Development Fund, Export-Import Bank of Korea à Cambodia
- Implementing Agency: ADB
- Executing Agency: Cambodian Ministry of Rural Development

(Asian Development Bank 2017b)

Building Resilience to Climate Change in Papua New Guinea (2013–on-going)

- Finance: $24.25 million
- Recipient country: Papua New Guinea
- Funding agency: Strategic Climate Fund
- Implementing Agency: ADB
- Executing Agency: Papua New Guinea Office of Climate Change and Development

(Asian Development Bank 2017c)

Promoting Urban Climate Change Resilience in Selected Asian Cities (2015–on-going)

- Finance: $15 million
- Recipient country: Regional
- Funding agency: Urban Climate Change Resilience Trust Fund under the Urban Financing Partnership Facility
- Implementing Agency: ADB
- Executing Agency: ADB

(Asian Development Bank 2017d)

project states that peer learning and joint training activities will be set up to build capacity across cities (ADB 2017c; ADB 2017d).

Among the 15 projects that had over $30 million of funding, monitoring and evaluation appeared to be very slim: the database contained no files in the 'evaluation documents' category for any of them. Although almost every project had a description of 'outputs', many of these outputs are subjective and vague, without explaining what specific targets are to be met or how they will be measured and monitored. Under the 'progress towards outcome' sections, many of the updates provided were about consultant packages being signed and managers being hiring, rather than climate-specific actions. These problems exist for other agencies, but the ADB site made some of these issues quite clear. Infrastructure construction displayed the most concrete output monitoring. For example, the project data sheet for the $59.75 million Cambodia Rural Roads Improvement Project states that 100 per cent of all road rehabilitation has been completed. Under the output of 'reduced vulnerability of project roads to climate change, however, progress reads merely as, 'Activities by the climate change adaptation consultants are ongoing.'

African Development Bank (AfDB)

The African Development Bank (AfDB) is a regional multilateral development finance institution with the aim of contributing to the economic development and social progress of the African countries that are members of the institution. It is the premier development finance institution on the continent with a central mission to reduce poverty, improve living conditions and mobilize resources for the continent's economic and social development. It promotes the investment of public and private capital in a myriad of sectors, including sustainable development or 'green growth'. The principal objective of the AfDB is to initiate sustainable economic development and social progress in its regional member countries with the central aim of poverty reduction. Two objectives to improve the quality of Africa's growth are: inclusive growth and the transition to green growth (African Development Bank 2012).

As a continent, Africa has contributed very little to global climate change and yet is threatened by the challenges it poses. The combination of geography, economic factors and its dependence on natural resources makes Africa the most vulnerable continent to the adverse effects of climate change. According to the AfDB's Climate Change Action Plan (CCAP), the negative effects of climate change have already reduced Africa's total GDP by 1.4 per cent and rising costs for adaptation to climate change are set to reach an annual 3 per cent of GDP by 2030 (African Development Bank 2011a). The CCAP focuses on three objectives: (1) low carbon development; (2) climate-resilient development; and (3) funding platforms. It intends to support African countries by strengthening their capacity to respond to climate change and capitalize on resources available from existing and proposed sources of climate finance. AfDB considers climate change an opportunity for Africa to adopt a development pathway that is 'climate-resilient and not carbon-intensive, that builds adaptive capacity and strengthens institutions' capability to integrate information into national planning and that strengthens national climate data systems'. The AfDB's approach to climate change is consistent with their commitment to 'green growth' (ibid.).

Our search of the AfDB Project Portfolio database with the key words 'capacity' and 'climate change' garnered 77 results with links to project pages; reviewing pages suggested only 18 projects were relevant for this analysis (Table 6.6). In all cases, it was unclear what amount of project

Table 6.6 Sample major capacity-building projects, African Development Bank

Institutional and Sustainability Support to Urban Water Supply and Sanitation Service Delivery (2016–on-going)

- Finance: $100,120,577
- Recipient country: Angola
- Funding agency: AfDB
- Implementing Agency: AfDB
- Executing Agency: Ministry of Energy and Water

(AFDB 2014)

Civil Society and Government Capacity Building within the Redd Framework (2012–on-going)

- Finance: $32,977,644.96
- Recipient country: République Démocratique du Congo, Equa
- Funding agency: AfDB (Co-Financer Delta)
- Implementing Agency: Woods Hole Research Center
- Executing Agency: AfDB

(AFDB 2010)

totals were allocated specifically to capacity building, and they were unclear on means of implementation.

Recipients of AfDB projects were either countries, several regions within countries or municipalities. There were no repeating implementing agencies in the data set, however, several projects included government agencies such as the Ministry of Agriculture or the Ministry of Water Resources as implementing agencies. In a few cases, international entities such as the Wildlife Conservation Society or the International Institute of Tropical Agriculture acted as implementing agencies.

The main objectives of the capacity building for climate change projects are consistent with the mission of the AfDB, with sustainable development as a central theme. Addressing the challenges of climate change among the many development challenges that plague the African continent, the projects in the dataset aim to 'strengthen the capacity of communities to address the interlinked challenges of adverse impacts of climate change, rural poverty, food security, and land degradation' (African Development Bank 2011b). Further, the majority of the projects list the enhancement of institutional and infrastructural capacity building as a central objective. Several projects focus on issues of water maintenance and sanitation, as desertification in the face of climate change is a challenge faced by many African nations. In the documentation we found online, the means of implementation for many of the projects was unclear and allocation of funding for capacity building was not delineated. The explanations of the means of implementation were often generalized, such as 'capacity building activities to enhance effectiveness of the agriculture sector delivery and project implementation'. Often the implementation was described as 'provided support'. In many cases, there was no description of implementation at all. However, the intentions for implementation were clear in the objectives of the majority of projects.

Inter-American Development Bank (IDB)

The Inter-American Development Bank (IDB) is committed to giving financial and technical support to countries in the Americas to achieve development in a 'sustainable, climate-friendly way' (IDB 2017). The IDB is the leading source of development in Latin America and the Caribbean, providing loans, grants and technical assistance for a variety of health, education and infrastructural projects. They are committed to achieving results in a measurable and transparent manner with high standards of integrity and accountability. The IDB focuses on three development challenges: (1) social inclusion and inequality; (2) productivity and innovation; and (3) economic integrity. These challenges include the three cross-cutting issues of gender equality and diversity, climate change and environmental sustainability, and institutional capacity.

The IDB includes a commitment to capacity building for climate change in its *Integrated Strategy for Climate Change Adaptation and Mitigation, and Sustainable and Renewable Energy* (IDB 2012). As a main objective, the strategy aims to provide guidance for dialogue between governments, civil society, and the private sector concerning climate policy agendas. Further, the IDB commits to integrating public and private financing and capacity building into a single framework for climate action. The IDB mentions the importance of capacity building in a myriad of sectors over 50 times. The Sustainable Energy and Climate Change Initiative was established with the specific intention of supporting knowledge and capacity building through the mainstreaming of climate change actions within IDB operational divisions and to 'build climate resilience in highly vulnerable sectors'. A commitment to capacity building is central to the mission of the IDB and is the principal objective of many of their funded projects in Latin America and the Caribbean.

In order to consider the role of capacity building for climate change in the IDB's funding allocation, we used the key words 'capacity building climate' in a search of the IDB's Project Database. The search produced 16 relevant projects, all with specific climate-related objectives (Table 6.7). The total funding for the projects was US$16.08 million with the largest single allocation of funding at US$5.94 million.

The IDB works through funding a stream directly with the national government of the recipient countries. Often, their funding is paired with additional funding provided by the national government for the projects. Therefore, many of the implementing agencies are either government entities or government-supported programmes. Of the 16 projects in the dataset, Mexico received funding for 3 projects and Guatemala for 2.

The importance of capacity building for climate change is transparent in the objectives of many IDB-funded projects. The majority of projects focus upon specific sectors of economic and developmental relevance with an

Table 6.7 Sample major capacity-building projects, Inter-American Development Bank

Institutional Strengthening in Support of Guyana LCDS (2012–on-going)
- US$ ($ cofinancing): $5.94 million
- Recipient country: Guyana
- Funding agency: IDB
- Implementing Agency: IDB
- Executing Agency: Office of the President, Guyana Forestry Commission

(IDB 2012)

Institutional Strengthening for the Implementation of the Energy Reform
- US$ ($ cofinancing): $1 million
- Recipient country: Mexico
- Funding agency: IDB
- Implementing Agency: IDB
- Executing Agency: División de Energía

(IDB 2017)

aim to strengthen each sector's resilience to climatic changes. These objectives include increasing the availability of information to address community vulnerabilities, strengthening management capabilities of specific government agencies, and building capacity in the agricultural sector. Further, several projects intend to scale up existing frameworks and programmes through 'knowledge, capacity building, and the dissemination of tools and platforms'. At a systemic level, several projects aim to build capacity through strengthening climate communications. For example, *Support to the Implementation of National Climate Change Strategy in El Salvador*, focuses on climate change adaptation policy through 'systems of monitoring, reporting, and verification for capacity building to climate change' (IDB 2014). Through the strengthening of communication regarding climate change, this project aims to increase access to vital information to communities most vulnerable to the effects of climate change. Finally, the IDB has funded three projects on a regional level specifically focused on capacity building in the Environment and Natural Disasters Sector. The projects aim to support capacity building, knowledge exchange and the economics of biodiversity and ecosystem services across Latin America and the Caribbean.

Each project in the dataset has clear goals for project implementation. The IDB is mindful of requesting 'counterpart funding' so the country itself is invested in the projects too. The bank tries to set countries up to attract more funding from other sources, including national government programmes and services. The implementing agencies in several cases are national research institutes or government agencies in charge of existing projects. This strategy for implementation requires the active participation of the national government of the recipient country in capacity building for climate change. Further, it aids the implementation of NAPAs and existing national programmes in need of additional funding and support.

The IDB includes clear objectives of capacity building for climate change in several of its projects. Each project in the dataset had clear objectives and means of implementation, specifically noting its intention to build capacity in the relevant sector. The use of national agencies as implementing agencies and existing projects as frameworks for funding creates a stricter outline for project implementation. This also increases opportunity for accountability, as the IDB can rely on existing structures within the agencies for reporting and evaluation. Further, the regional specificity creates greater understanding of the social and environmental needs of each recipient country. For example, the IDB is able to focus on small island nation issues specific to the Caribbean. Finally, the IDB provides clear objectives and means of implementation for each of its projects, but does not include in-depth project reviews.

The World Bank

The World Bank is a global leader in financial and technical assistance to developing countries. It is a unique partnership of five institutions committed to two development goals to achieve by 2030: (1) end extreme poverty by decreasing the percentage of people living on less than $1.90 a day to no more than 3 per cent; and (2) promote shared prosperity by fostering the income growth of the bottom 40 per cent for every country. They provide low-interest loans, zero to low-interest credits, and grants to developing countries along with fostering trust fund partnerships with bilateral and multilateral donors (World Bank 2017).

The World Bank is a leading contributor to climate finance, recognizing the glaring intersection of global poverty and global environmental change. In the wake of the Paris Agreement at COP21, the World Bank published their Climate Change Action Plan for 2016–2020 (World Bank 2016). The Action Plan commits to engage in financial and technical assistance to 'feed nine billion people by 2050, provide affordable energy access to all, and extend housing and services to two billion new urban dwellers–and to do so while minimizing emissions and boosting resilience'. The World Bank aims to scale up climate action, integrate climate change across its operations, and work more closely with other organizations to achieve its goals. In order to reach impact at scale, the Climate Action Plan is focused on shaping national investment plans and policies, through finance streams to national governments. The World Bank commits to increasing the climate-related share of its portfolio from 21 to 28 per cent by 2020 in order to respond to client demand with total financing reaching $29 billion per year by 2020.

The World Bank intends to achieve these proposals through strategic shifts in climate action. In implementation, the World Bank seeks to accelerate support for countries and companies to implement the plans they have developed. On 'convergence', the Bank says its climate and development agendas will be fully integrated into strategies and operations and global and country-level action will be aligned. To maximize impact, the Bank works to increase its focus on impact at scale, including shaping national investment policies and programmes and mobilizing private finance. On resilience, the Bank is working to rebalance its portfolio, putting greater focus on adaptation and resilience. Finally, the Bank recognizes that reaching global climate commitments will require a shift from business as usual, a transformation.

Each of these strategic shifts suggests the need for substantial capacity building at the national level. Further, the World Bank says it will support countries in translating their NDCs into climate policies and climate-smart investment plans. They will support the mainstreaming of climate considerations into national and local policies. Finally, the World Bank commits to scale up in renewable energy and energy efficiency, sustainable mobility, sustainable resilient cities, and climate-smart land use, water and food security (World Bank 2016c).

In order to gain a broad picture of the World Bank's contribution to capacity building programmes, we ran a search of the World Bank Project and Operations Database. We used the key words 'capacity building' and 'climate', sorting results chronologically. This search returned over 1,000 results. To create a dataset, we chose no more than 3 projects from each of the 20 pages of results based on the project's relevance to climate change and capacity building. In addition, we considered geographic diversity in our choice of which projects to analyse. Our search produced 43 projects for analysis (Table 6.8).

Of the 43 projects in the sample, 19 provide funding through the International Development Association. The recipients of this funding are entirely encompassed by various national governments for large-scale capacity building projects. Similarly, six projects provided funding through the International Bank for Reconstruction and Development to national government recipients. Other agencies such as the Strategic Climate Fund Grant provided funding to departments within national governments such as the Ministry of Environment and Sustainable Development of Burkina Faso. Only eight projects provided funding to agencies within national governments and only one provided funding to a non-governmental organization.

The most consistent goals among the World Bank-funded projects were building climate-related disaster resilience for vulnerable communities and infrastructure management of water systems. Included in these goals were intentions of more reliable meteorological information systems with speedy and efficient responses to climate changes, sustainable land management for agro-ecosystems, and the development of approaches that would enable targeted communities to adapt to the potential impacts of climate variability and change. One project, the *Higher Education Quality Enhancement Project* in Bangladesh, focuses on the power of the education sector by

Table 6.8 Sample major capacity-building projects, the World Bank

Mekong Delta Integrated Climate Resilience and Sustainable Livelihoods Project (2016–2022)
- US$ ($ cofinancing): $310 million
- Recipient country: Vietnam
- Funding agency: World Bank
- Implementing Agency: International Development Association
- Executing Agency: Ministry of Natural Resources and the Environment

(World Bank 2016a)

Coastal Region Water Security and Climate Resilience Project (2014–2021)
- US$ ($ cofinancing): $200 million
- Recipient country: Kenya
- Funding agency: World Bank
- Implementing Agency: International Development Association
- Executing Agency: National Treasury

(World Bank 2014)

Integrated Disaster Risk Management and Resilience Program (2016–2021)
- US$ ($ cofinancing): $200 million
- Recipient country: Morocco
- Funding agency: World Bank
- Implementing Agency: International Bank for Reconstruction and Development (World Bank)
- Executing Agency: Government of Morocco

(World Bank 2016b)

PK Sindh Irrigated Agriculture Productivity Enhancement Project (2015–2021)
- US$ ($ cofinancing): $187 million
- Recipient country: Pakistan
- Funding agency: World Bank
- Implementing Agency: International Development Association
- Executing Agency: Sindh Agricultural Department

(World Bank 2015)

Kenya Water Security and Climate Resilience Project (2013–2022)
- US$ ($ cofinancing): $182.670 million
- Recipient country: Kenya
- Funding agency: World Bank
- Implementing Agency: International Development Association, Germany
- Executing Agency: National Treasury of Kenya

(World Bank 2013)

aiming to improve the quality and relevance of the teaching and research environment in higher education institutions (World Bank 2009). The project intends to encourage both innovation and accountability within universities by enhancing the technical and institutional capacity of the higher education sector. Consistently, the World Bank seems to fund projects through national and regional governments that target the most important elements of building climate-resilient communities.

The World Bank funding recipients sought to meet these goals through clear and tangible implementation strategies. Several projects focused on the built environment, intending to strengthen infrastructural resilience to climate change though the enforcement and construction of public projects such as dams and irrigation systems. Further, several projects provided provisions for training community members and officials in flood mitigation, disaster resilience, and natural resource management. One project focused on the mainstreaming of climate change through climate communication on Reducing Emission from Deforestation and Forest Degradation or REDD+. Finally, funding was provided to vulnerable communities for both educational and structural capacity building at a local level.

The sheer number of projects funded by the World Bank with relevance to climate change and capacity building provided ample opportunities for analysis. The capacity-building elements of the analysed projects were very clear with translatable and transparent intentions for project implementation. Though most of the projects provided funding directly to national governments, the projects were often carried out by smaller government departments and local communities. Most of the projects are on-going with little data available regarding the results of the intended implementation. With this information not yet complete, understanding the breadth of the capacity actually being built is difficult. However, the clear objectives and means of implementation delineated by most of the projects predict similar transparency from the reporting of the results.

The Pilot Program for Climate Resilience (PPCR)

The Pilot Program for Climate Resilience (PPCR) is a funding window for the Climate Investment Funds used primarily for climate change adaptation and resilience building projects. The Climate Investment Funds (CIF), comprised of four programmes, including the PPCR, is a $8.3 billion fund that provides 72 countries with resources to manage the challenges of climate change and mitigate its effects through the reduction of greenhouse gas emissions. The PPCR funds technical assistance and investments to 'support countries' efforts to integrate climate risk and resilience into core development planning and implementation' (CIF 2017). The PPCR uses a two-phase programmatic approach: assisting national governments in integrating climate resilience into development planning, and providing additional funding to put the plan into action and piloting innovative public and private sector solutions to essential climate-related challenges.

To examine the role of capacity building for climate change in the PPCR-funded projects, we used the CIF project database and filtered the search for only PPCR projects. This produced 78 projects (Table 6.9). The keyword search 'capacity building' produced no results, so we searched through individual project documents for its mention. This reduced the dataset to 41 with a total of $591.6 million in CIF funding and $958.06 million in cofinancing.

Table 6.9 Sample major capacity-building projects, PPCR

Climate Resilience: Integrated Basin Management Project (2013–on-going)

- US$ ($ cofinancing): $46 million ($9 million)
- Recipient country: Bolivia
- Funding agency: PPCR
- Implementing Agency: IBRD
- Executing Agency: PPCR Coordination Unit (UCP-PPCR)

(CIF 2013a)

Strengthening Climate Resilience in the Kafue River Basin (2013–on-going)

- US$ ($ cofinancing): $39 million ($40 million)
- Recipient country: Zambia
- Funding agency: PPCR
- Implementing Agency: AfDB
- Executing Agency: National Climate Change Secretariat

(CIF 2013b)

The PPCR funds projects in conjunction with several multilateral development banks (MDBs). For the projects specific to capacity building for climate change, the International Bank for Reconstruction and Development, i.e. the World Bank, was the most significant MDB partner, supporting 15 of the projects. The Asian Development Bank (ADB) supported 12 projects. Several other MDBs contributed to projects in the data set: the International Finance Corporation, the Inter-American Development Bank, the African Development Bank, and the European Bank for Reconstruction and Development. In partnership with these co-financing multilateral development banks, government agencies within the recipient countries carried out the implementation of project components.

The PPCR focuses its capacity-building initiatives on implementation for adaptation measures. Each relevant project delineates a clear objective with steps for implementation. For projects focused on development and transfer of technology, aims included 'provision of technical assistance for the purpose of integrating climate change adaptation and disaster risk reduction into programs, plans, policies and projects' (CIF 2012). Institutional capacity building, including the strengthening or establishment of national climate change secretariats or national focal points, involved, 'evidence of strengthened government capacity and coordination mechanism to mainstream climate resilience'. CIF projects had budgets ranging between $10–45 million, around $15 million per project average; with cofinancing, these can be substantially higher. The percentage of total project budget that was earmarked for capacity building varied highly, ranging from 0.6 per cent to 100 per cent and everywhere in between. Many projects are almost entirely capacity building, such as institutional strengthening of the hydro-meteorological services and development of data management tools = 100 per cent, while many other projects had capacity building tacked on (in the 1–5 per cent range). Projects were located in a mix of LDCs, SIDS, and Africa.

Capacity building for climate change is a core objective of the majority of the PPCR-funded projects. Examples include a feasibility study for low-cost climate-resilient housing, making climate information available to users, building capacity of hydrometeorological services, promoting climate-resilient agriculture, improving quality and use of climate-related data for effective planning and action, and mainstreaming climate adaptation in urban planning.

Overall, then, we see that the MDBs are rapidly increasing funding to climate change, including ambitious targets for proportions of their portfolios that are required to be focused on this problem in the next few years. Capacity building was seen to be a frequent theme of their work, but the agencies significantly varied in their approaches to doing so.

Capacity-building initiatives under the UNFCCC thematic bodies and funds

To gain a broad perspective on how UN agencies were taking up capacity building in their projects and programmes, we reviewed databases of the Global Environment Facility (the GEF), its co-implementers the UNDP and the UNEP, and reviewed project lists from the Adaptation Fund of the UNFCCC.

Global Environment Facility (GEF)

The Global Environment Facility (GEF) is a 'catalyst for action on the environment'. Established in preparation for the 1992 Rio Earth Summit, the GEF has been a leader in climate finance on a global scale since its initiation (UNEP 2017a). The GEF reports that it operates with global agencies to target the most challenging environmental issues with strategic investments. The GEF is a partnership of 18 agencies, including the UN agencies described in this section, along with several multilateral development banks mentioned in this chapter and some key international NGOs. The GEF works with 183 countries in a network of civil society, private organizations, and government agencies. It operates as a financial mechanism for five international environmental conventions, including the United Nations Framework Convention on Climate Change (UNFCCC).

The GEF administers several trust funds for the UN climate convention, including the Special Climate Change Fund (SCCF) and the Least Developed Countries Fund (LDCF), and provides secretariat services to the Adaptation Fund (AF). The GEF Trust Fund provides funding for developing countries and countries with transitioning economies for the purpose of meeting the objectives and obligations of the international environmental conventions and agreements. Financial contributions are replenished every four years and are provided by 39 donor countries.

The Least Developed Countries Fund (LDCF) operates under the UNFCCC, addressing the needs of the 52 Least Developed Countries (LDCs) and their specific vulnerabilities to the adverse impacts of climate change. The LDCF addresses vulnerability in sectors that are central to development and livelihoods, including: water, agriculture and food security; health, disaster risk management and prevention; infrastructure; and fragile ecosystems. The LDCF is the only existing fund tasked to finance the preparation and implementation of National Adaptation Programmes of Action (NAPAs). It is also aimed at financing the entire cost of adaptation for NAPA projects, but significant shortfalls exist on reaching that goal. Specifically, it finances 'urgent and immediate adaptation actions that reduce vulnerability and increase adaptive capacity to the impacts of climate change'. The LDCF and the GEF Trust Fund, and occasionally the SCCF, are the primary funding sources for the projects of the United Nations Development Programme (UNDP) and the United Nations Environment Programme (UNEP) focused on capacity building for climate change (Biagini & Dobardzic 2011). The following section considers the role of capacity building for climate change in projects funded primarily by the GEF Trust Fund and the LDCF and carried out by the UNDP and UNEP. Further, it considers projects funded by the Adaptation Fund, implemented by a myriad of agencies.

The United Nations Development Programme (UNDP)

The United Nations Development Programme (UNDP) is the global development network for the United Nations. Working in 170 countries, the UNDP focuses on the eradication of poverty and the reduction of inequalities at a global scale (UNDP 2017b). The UNDP prioritizes building solutions in three main areas: (1) sustainable development; (2) democratic governance and peacebuilding; and (3) climate and disaster resilience. In regard to the latter, the UNDP recognizes the exacerbating effects of climate change upon existing economic, political and humanitarian stresses and its effect on development expectations and programmes. The UNDP works with planning bodies at national, regional and local scales to build capacity to effectively respond to climate change and promote low-emission, climate-resilient development. The UNDP addresses climate-smart development through several objectives: connecting countries to knowledge and resources, building more resilient societies, and 'strengthening the capacity of countries to access, manage, and account for climate finance'. The UNDP operates as an implementing agency for other UN-funded agencies such as the Least Developed Countries Fund and the Global Environment Facility.

In order to examine the UNDP projects relevant to capacity building for climate change, we used the Global Environment Facility project database. Using the key words 'capacity building' and 'climate change', specifying the agency as the UNDP and the focal area as climate change, and setting the timeframe to projects initiated after 2013, our search produced 37 projects (Table 6.10). Of these projects, all but two were funded in part by either the GEF Trust Fund of the LDCF. Eleven projects included support to countries in creating their update reports for the UNFCCC. Financing from the GEF totalled $140,612,636 with $678,959,139 in cofinancing from other agencies.

As previously mentioned, the GEF Trust Fund and the LDCF are the funding agencies of all but two projects in the dataset, which are funded by the Special Climate Change Fund. The UNDP, working with government agencies and local organizations, is the implementing agency for all the projects. The UNDP provides support to the government agency in charge of the implementation of the project. These agencies are often ministries of environment, conservation, or natural resources. Because each project focuses on only one country, the UNDP works with the national agency most relevant to the project objectives. Only one project, *Building Resilience of Health Systems in Asian LDCS to Climate Change*, was not paired with a government agency and instead provided support to the World Health Organization (WHO) and operates in several countries ('Building resilience of health' 2017).

Several of the projects operate with the sole objective of assisting the preparation of Update Reports for the fulfilment of obligations under the United Nations Framework Convention on Climate Change. Each project provides support to one country in the development of communication strategies and implementation of obligations through funding from the GEF Trust Fund. Several projects support capacity in specific economic and environmental sectors, such as agriculture and water management.

Table 6.10 Sample major capacity-building projects, UNDP

Adapting to Climate Change Induced Coastal Risks Management in Sierra Leone (2014–2016)

- US$ ($ cofinancing): $9,975,000 ($30,000,000)
- Recipient country: Sierra Leone
- Funding agency: LDCF
- Implementing Agency: UNDP
- Executing Agency: Environment Protection Agency Sierra Leone

(GEF 2016)

Mainstreaming Climate Risk Considerations in Food Security and IWRM in Tsilima Plains and Upper Catchment Area (2013–2015)

- US$ ($ cofinancing): $9.05 million ($27.5 million)
- Recipient country: Eritrea
- Funding agency: LDCF
- Implementing Agency: UNDP
- Executing Agency: Ministry of Land, Water and Environment, Eritrea

(GEF 2015)

These projects focus on one specific sector, and often one geographic region, and address challenges exacerbated by climate change. Further, several projects support economic infrastructure for development through protection from climate-related hazards. Other objectives include mainstreaming adaptation to climate change into energy policies and management strategies. This entails working with government agencies to implement climate-focused development into their main programming.

As mentioned previously, each project carries out implementation in conjunction with government agencies. These agencies range from ministries specific to climate change to ones focused on broader development sectors, such as disaster management. Many, but not all, of the projects included detailed project documents outlining the means of implementation. These means included direct integration of development techniques, such as water harvesting, developed at local and national research institutions. Similarly, they included the implementation of innovative and sustainable economic instruments established by government agencies for widespread development. As the UNDP prioritizes addressing the intersection of climate change in development issues, much of the implementation carried out combines development projects with capacity building for climate change.

The UNDP approaches challenges of climate change through the lens of development, always including development strategies in projects for capacity building for climate change. This cross-cutting approach creates a greater opportunity for real capacity to be built, as it addresses development challenges that must take the present and future challenges of a changing climate into consideration. As few projects in this dataset have clear results to analyse, this is simply conjecture. However, the majority of projects in the database are accompanied by clear project documents including Project Identification Forms (PIF) that contain detailed assessments of the project purpose, funding allocations, and means of implementation.

The United Nations Environmental Programme (UNEP)

The United Nations Environmental Programme (UNEP) is the central agency of the United Nations coordinating environmental activities, supporting developing countries in implementing sustainable development practices. UNEP has been a leader in supporting projects and providing scientific guidance in the field of climate change, including aid in the facilitation of 15 National Adaptation Plans of Action (NAPAS). It has implemented or is in the process of implementing over 80 adaptation projects at global, regional and national levels. The stated top objective of UNEP in these adaptation projects is 'building capacity of stakeholders', particularly in ecosystem management and sustainable development. UNEP approaches the challenges of climate change adaptation with three main focuses:

(1) science and assessment; (2) knowledge and policy support; and (3) building the resilience of ecosystems for adaptation. Central to UNEP's adaptation work is the EbA flagship programme which aims to build resilience to climate change through the restoration of key ecosystems such as river basins, mountains, coastal zones and drylands. This programme encompasses the objectives of national, regional, and local adaptation projects funded by the UNEP (UNEP 2015).

Our search of the GEF project database using the key words 'capacity building' and 'climate change', narrowed to 'climate change' focal area and UNEP, produced nine projects (Table 6.11). Funding from the UNEP for these projects totalled $21,804,500 with $71,832,912 in cofinancing from other agencies.

The majority of these projects used government departments and ministries as the main implementing agencies. Often these departments focused on energy, environment, or science. In addition, several projects included research institutions (in the Global North) such as the World Resource Institute and the Stockholm Environment Institute as implementing agencies. Each project executive summary included the names of all secondary

Table 6.11 Sample major capacity-building projects, UNEP

Adapting Coastal Zone Management to Climate Change in Madagascar Considering Ecosystem and Livelihoods (2011–2014)

- $5,337,500 ($12,050,000 cofinancing)
- Recipient country: Madagascar
- Funding agency: LDCF
- Implementing Agency: UNEP
- Executing Agency: Madagascar (Ministry of Environment and Forests)

(GEF 2017a)

Enhancing Capacity, Knowledge and Technology Support to Build Climate Resilience of Vulnerable Developing Countries (2012–2017)

- $4,900,000 ($34,700,000 cofinancing)
- Recipient Countries: Seychelles, Nepal, Mauritania
- Funding Agency: Special Climate Change Fund
- Implementing Agency: UNEP
- Executing Agency: China National Development and Reform Commission (NDRC), African Climate Policy Centre (ACPC), Institute of Geographical Science and Natural resource Research, Chinese Academy of Sciences (IGSNRR-CAS) and Stockholm Environment Institute (SEI)

(GEF 2017b)

Generation and Delivery of Renewable Energy Based Modern Energy Services in Cuba; the case of Isla de la Juventud

- $5,337,500 ($12,050,000 cofinancing)
- Recipient country: Madagascar
- Funding agency: LDCF
- Implementing Agency: UNEP
- Executing Agency: Madagascar (Ministry of Environment and Forests)

(GEF 2017c)

agencies receiving funding from the main implementing agency for specific aspects of the project. In some cases these were private members of the energy sector or sub-divisions of national or regional government.

Capacity building is sought in a myriad of sectors and through a variety of activities. For example, the project *Capacity Building Programme to Implement South Africa's Climate National System*, aims to 'strengthen South Africa's capacity to meet the transparency requirement of the Paris Agreement' through several actions such as the support of transformational shifts towards low-emission and resilient development paths (GEF 2017d). Other projects include goals for building capacity in the monitoring and predicting of climate change impacts through early warning systems for extreme weather events. The main of objective of the project *Enhancing Capacity, Knowledge and Technology Support to Build Climate Resilience of Vulnerable Developing Countries*, is to 'build climate resilience using Ecosystem Based approaches to Adaptation (EBA) through capacity building, knowledge support and concrete, on-the-ground interventions' (GEF 2017b). The majority of the projects sought implementation through strengthening of aspects of the information and technology sectors. Activities of one project included local consultations, workshops and meetings, and institutional and financial strategy.

The Adaptation Fund (AF)

The Adaptation Fund was established under the Kyoto Protocol of the UNFCCC in 1998 with a commitment to financing projects and programmes for the adaptation needs of vulnerable communities in developing countries. All initiatives address climate change through a framework of resiliency and capacity building. Its Readiness Programme for Climate Finance aims to help strengthen the capacity of national and regional entities to receive and manage climate financing as they continue to build resilience to the changing climate across all sectors (Adaptation Fund 2014).

We searched the AF's 63 Projects and Programmes; 7 directly mention capacity building. 4 of the 6 projects were implemented in India by the agency NABARD. 2 of the projects were implemented by other UN agencies, IFAD and UNDP (Table 6.12). Total AF funding for capacity building projects was $36,886,947. The main implementing agency for the capacity-building projects was India's National Bank for Agriculture and Rural Development, the International Fund for Agricultural Development (IFAD) and UNDP. In addition to the main implementing agencies, each project delineated local leaders, technical experts, and educational professionals who supported the project through specific programme implementation.

Every AF project in the dataset mentions building the adaptive capacity of a specific sector as its central goal. These sectors include small-scale family agricultural producers, fishermen, and communities depending on

Table 6.12 Sample major capacity-building projects, the Adaptation Fund

Integrated Programme to Build Resilience to Climate Change and Adaptive Capacity of Vulnerable Communities in Kenya (2014–on-going)
- Finance: $9,998,302
- Recipient country: Kenya
- Funding agency: AF
- Implementing Agency: NEMA (National Environment Management Authority)
- Executing Agency: Kenya Forestry Research Institute, Tana and Athi Rivers Development Authority, Coast Development Authority

(Adaptation Fund 2015a)

Climate Smart Agriculture: Enhancing Adaptive Capacity of the Rural Communities in Lebanon (2012–2016)
- Finance: $7,860,825
- Recipient country: Lebanon
- Funding agency: AF
- Implementing Agency: IFAD (International Fund for Agricultural Development)
- Executing Agency: Ministry of Agriculture

(Adaptation Fund 2015b)

Enhancing Adaptive Capacity of Communities to Climate Change-Related Floods in the North Coast and Islands Region of Papua New Guinea (2012–2016)
- Finance: $$36,886,947
- Recipient country: Papua New Guinea
- Funding agency: AF
- Implementing Agency: UNDP
- Executing Agency: Office of Climate Change and Development

(Adaptation Fund 2015c)

natural resource management. *Integrated Programme to Build Resilience to Climate Change and Adaptive Capacity of Vulnerable Communities in Kenya* provided funding for the implementation of a ‘Climate Change Observatory’, or a group of invited members of universities, climate change units, and experts from meteorological departments with the intended role of expert data analysis (Adaptation Fund 2015a). With its focus on supporting adaptive capacity, capacity building is central to many of the projects. The transparency of the Adaptation Fund project database is unparalleled; with multi-page project descriptions with detailed plans for implementation, distinct allocation of actors roles, and clear budget proposals, the Adaptation Fund Project provide ample opportunity for high accountability, monitoring and evaluation. This degree of accountability creates structures for legitimate capacity-building projects with tangible results and significant impact.

Conclusion: varying approaches, disparate transparency

Our overall observation is that agencies are quite uneven in their attention to capacity building in the area of climate change management. The language is quite varied in how agencies discuss this work: some use terms like adaptive capacity, while others discuss resilience. We noticed some areas that are underdeveloped in these portfolios. Especially for the focus of this book, education and training to address climate change science and management were the focus only for NORAD projects. Knowledge transfer seems more enduring if people are being trained in this area, especially in-country, in university or ad hoc training programmes. Expanding and strengthening regional and local knowledge systems and practices could be done with support for local community organizations and municipal governments. Getting this support to local people and building sustainable training programmes could be a major focus of these efforts. Building the 'soft infrastructure' of educational capacity should be considered along with building sea-walls and breeding drought-resistant crops.

Of the 15 UNFCCC (Decision 2/CP.7) Capacity Building themes laid out in 2001, institutional capacity building was consistently mentioned, and the implementation of adaptation measures (UNDP and GEF projects, especially), and the efficiency of national communications was often mentioned, especially in disaster awareness and warnings to vulnerable populations. 'Education, training and public awareness' were not as common as the other priority areas, along with 'research and systematic observation, including meteorological, hydrological and climatological services'.

In reviewing capacity-building projects from these agencies, capacity-building efforts seemed to line up with the larger focus of each agency. For example, UNDP capacity-building projects tended to focus on the intersection of development and environment, and on how climate change exacerbates issues of economic well-being. Similarly, the UNEP focuses on ecosystem issues, so their projects focused more on how climate change affects ecosystems. The impact of climate change on fisheries and agriculture is often seen in the context of how they are supported by aquatic and terrestrial ecosystems. The African Development Bank was unique in focusing on attempting to bring Africa into inclusion in economic growth, and considering that Africa is the least responsible and most vulnerable to climate change. The AfDB seeks to have the region included in growth along developmental pathways that are 'climate resilient and less carbon intensive' and that build adaptive capacity and strengthen institutions (African Development Bank 2012). Many of their projects focused on hiring climate change professionals and training others in national and municipal bureaucracies on issues of climate change. The IDB focused on cross-cutting issues like gender equality and diversity, climate change and environmental sustainability, and institutional capacity. On climate change, they sought to provide guidance in dialogues between governments, civil society and the private sector concerning the climate change policy agenda. In Mexico, the IDB sought to strengthen institutional capacity for the implementation of energy reform, working in one project to support the 'regulatory framework and institutional capacity' in energy reform, including for renewable energy in the private sector.

Often bilateral agencies were consistent with national agendas, for example, CIDA focused strongly on women and gender issues, including when working on climate change and building capacity in developing countries. UK Aid/DFID is focused on poverty reduction, and this shaped their climate change and capacity-building work. As mentioned above, the USAID information now entirely excludes mention of climate change at the level of the website, but individual projects did have climate-focused capacity building during the Obama Administration. NORAD was extraordinary in that all its projects were university-based, aimed at building staff capacity by creating and expanding Master's and PhD programmes in developing countries that focused on climate change research and management. 'Strengthening education and research capacity in climate change and natural resource management in universities', as described in a project on fisheries and aquaculture management in Sri Lanka and Vietnam (NORAD 2013b), is precisely the approach we will be advocating for in later chapters in this volume.

On transparency and accountability, there was huge variation by agency, even from one project to another within a single agency, and even those reported in the same year. For example, the GEF has detailed project documents for many projects, but others had only one letter of support. Still, our review showed the GEF is highly transparent, in line with other World Bank-administered programmes. (The World Bank was the target of decades of organizing for greater transparency and accountability, and has made great progress in that regard.) SIDA projects were thin on documentation that we could locate, while NORAD had detailed information available, including a list of key contact people with their emails and links to news coverage of projects and institutional web links, but they lacked detailed project description documents. The IDB had excellent documentation, including information on allocation of funds, mapping, project documents, supporting letters from host governments. The USAID website is deeply problematic at present, since in 2017 there is no longer any climate change sector listed and it is difficult to locate projects dealing with the issue. Therefore, we were forced to retrieve information from the OECD Creditor Reporting System database. Therefore, it is difficult to assess whether the US results can be seen as parallel to others in this round-up. This is ironic, given the USAID website states that 'The U.S. Agency for International Development is committed to the President's Open Government initiative, upholding the values of transparency, participation, and collaboration in tangible ways that benefit the American people.' The Adaptation Fund was consistent in providing project documents, and also usefully provided a clear breakdown of budget components and objectives. Again, any World Bank-administered agency (LDCF, Adaptation Fund, GEF) had the best transparency.

References

Adaptation Fund. (2014). "Adaptation Fund Convenes Practitioners & Experts For Climate Finance Capacity Building" Press Release. Available at www.adaptation-fund.org/adaptation-fund-convenes-practitioners-experts-for-climate-finance-capacity-building/ (accessed 30 December 2017).

Adaptation Fund. (2015a). Integrated programme to build resilience to climate change and adaptive capacity of vulnerable communities in Kenya. Available at: www.adaptation-fund.org/project/integrated-programme-to-build-resilience-to-climate-change-adaptive-capacity-of-vulnerable-communities-in-kenya/

Adaptation Fund. (2015b). Climate smart agriculture: enhancing adaptive capacity of the rural communities in Lebanon (AgriCAL). Available at: www.adaptation-fund.org/project/climate-smart-agriculture-enhancing-adaptive-capacity-of-the-rural-communities-in-lebanon-agrical/

Adaptation Fund. (2015c). Enhancing adaptive capacity of communities to climate change-related floods in the north coast and islands region of Papua New Guinea. www.adaptation-fund.org/project/enhancing-adaptive-capacity-of-communities-to-climate-change-related-floods-in-the-north-coast-and-islands-region-of-papua-new-guinea/

Adaptation Fund. (n,d,). Readiness programme for climate finance. Available at; www.adaptation-fund.org/readiness/ (accessed 18 December 2017).

AdaptationWatch. (2016). *AdaptationWatch*. Available at: www.adaptationwatch.org/#about (accessed 19 September 2017).

ADB. (2017a). Regional: Enhancing Gender Equality Results in the Southeast Asian DMCs. Asian Bank Development. Available at: www.adb.org/projects/49396-001/main.

ADB. (2017b). Regional: Targeted Capacity Building for Mainstreaming Indigenous Peoples Concerns in Development. Available at https://adb.org/projects/documents/targeted-capacity-building-mainstreaming-indigenous-peoples-concerns-development-tcr (accessed 30 December 2017).

ADB. (2017c). Tajikistan: Building Capacity for Climate Resilience. Available at: www. adb.org/projects/45436-001/main.

ADB. (2017d). Regional: Harnessing Climate Mitigation Initiatives to Benefit Women. Available at www.abd.org/projects/45039-001/main (accessed 30 December 2017).

African Development Bank. (2010). *Civil society and government capacity building within the REDD framework*. Available at: www.afdb.org/en/projects-and-operations/project-portfolio/p-z1-c00-029/ (accessed 21 September 2017).

African Development Bank. (2011a, November). *Climate Change, Gender, and Development in Africa* (Vol. 1, Issue 1). Available at: www.afdb.org/fileadmin/uploads/afdb/Documents/Publications/Climate%20Change%20Gender%20and%20Development%20in%20Africa.pdf (accessed 21 September 2017).

African Development Bank. (2011b). *Climate Change Action Plan 2011–2015*. Available at: www.afdb.org/fileadmin/uploads/afdb/Documents/Policy-Documents/Climate%20Change%20Action%20Plan%20%28CCAP%29%202011-2015.pdf (accessed 21 September 2017).

African Development Bank. (2012). *Solutions for a changing climate*. Available at: www.afdb.org/fileadmin/uploads/afdb/Documents/Generic-Documents/The%20Solutions%20for%20a%20Changing%20Climate%20The%20African%20

Development%20Bank%27s%20Response%20to%20Impacts%20in%20 Africa.pdf (accessed 21 September 2017).

African Development Bank. (2014). *Institutional and sustainability support to urban water supply and sanitation service delivery*. Available at: www.afdb.org/en/projects-and-operations/project-portfolio/p-ao-e00-005/ (accessed 21 September 2017).

Asian Development Bank. (2017a). Bangladesh: Coastal Climate-Resilient Infrastructure Project. Available at: www.adb.org/projects/45084-002/main#project-pds

Asian Development Bank. (2017b). Cambodia: Rural Roads Improvement Project. Available at: www.adb.org/projects/42334-013/main#project-pds

Asian Development Bank. (2017c). Papua New Guinea: Building resilience to climate change in Papua New Guinea. Available at: www.adb.org/projects/46495-002/main#project-overview

Asian Development Bank. (2017d). Regional: Promoting urban climate change resilience in selected Asian cities. Available at: www.adb.org/projects/48317-001/main#project-overview

Asian Development Bank. (2017e). ADB's Focus on Environment. Available at: www.adb.org/themes/environment/main (accessed 18 December 2017).

Biagini, B., & Dobardzic, S. (2011). *Accessing resources under the Least Developed Countries Fund*. Washington, DC: World Bank.

Chiang, K. K. (2008). Building climate change research capacity in Africa and Asia. The Stockholm Environmental Institute, Asia. Available at: www.sei-international.org/mediamanager/documents/Publications/Climate/sarec081124.pdf (accessed 18 December 2017).

CIDA. (2017). CIDA's Children and Youth Strategy. Available at: http://international.gc.ca/world-monde/issues_development-enjeux_developpement/priorities-priorites/children_youth_strategy-strategie_enfants_jeunes.aspx?lang=eng

CIF. (2012). Mainstreaming climate resilience into development planning. Available at: www.climateinvestmentfunds.org/projects/mainstreaming-climate-resilience-development-planning (accessed 18 December 2017).

CIF. (2013a). Climate resilience – Integrated Basin Management Project. Available at: www.climateinvestmentfunds.org/projects/climate-resilience-integrated-basin-management-project

CIF. (2013b). Strengthening climate resilience in the Kafue River Basin. Available at: https://climateinvestmentfunds.org/projects/strengthening-climate-resilience-kafue-river-basin

CIF. (2017). What we do. Available at: www.climateinvestmentfunds.org/about (accessed 18 December 2017).

DECC. (2016, 10 October). International carbon capture and storage capacity building. In Development Tracker. Available at: https://devtracker.dfid.gov.uk/projects/GB-4-91089 (accessed 19 September 2017).

DFID. (2017a, June 30). Lower Limpopo River Valley flood reconstruction and climate resilience project. In Development Tracker. Available at: https://devtracker.dfid.gov.uk/projects/GB-1-204064 (accessed 19 September 2017).

DFID. (2017b, August 21). Forest governance, markets and climate. In Development Tracker. Available at: https://devtracker.dfid.gov.uk/projects/GB-1-201724 (accessed 19 September 2017).

DFID. (2017c). About us. Available at: www.gov.uk/government/organisations/department-for-international-development/about

ECBI. (2014). Proposal for ECBI Phase 4. Available at: www.eurocapacity.org/downloads/ecbi_Phase_IV_Proposal.pdf (accessed 30 December 2017).

Global Affairs Canada. (2016–2017). *Report on plans and priorities 2016–17*. Available at: http://international.gc.ca/gac-amc/assets/pdfs/publications/plans/rpp/RPP_2016_2017_ENG.pdf (accessed 19 September 2017).

Global Affairs Canada. (2017a, September 19). *Project profile: Energy sector capacity building*. Available at: http://w05.international.gc.ca/projectbrowser-banqueprojets/project-projet/details/A035511001 (accessed 19 September 2017.

Global Affairs Canada. (2017b, September 19). *Project profile: World Bank Bio-Carbon Plus Fund*. Available at: http://w05.international.gc.ca/projectbrowser-banqueprojets/project-projet/details/M013504001 (accessed 19 September 2017).

GEF. (2015). Mainstreaming climate risk considerations in food security and IWRM in Tsilima Plains and Upper Catchment Area. Available at: www.thegef.org/project/mainstreaming-climate-risk-considerations-food-security-and-iwrm-tsilima-plains-and-upper

GEF. (2016). Adapting to climate change-induced coastal risks management in Sierra Leone. Available at: www.thegef.org/project/adapting-climate-change-induced-coastal-risks-management-sierra-leone

GEF. (2017a). Adapting coastal zone management to climate change in Madagascar considering ecosystem and livelihoods. Available at: www.thegef.org/project/adapting-coastal-zone-management-climate-change-madagascar-considering-ecosystem-and

GEF. (2017b). Enhancing capacity, knowledge and technology support to build climate resilience of vulnerable developing countries. Available at: www.thegef.org/project/enhancing-capacity-knowledge-and-technology-support-build-climate-resilience-vulnerable

GEF. (2017c). Generation and delivery of renewable energy-based modern energy services in Cuba; the case of Isla de la Juventud. Available at: www.thegef.org/project/generation-and-delivery-renewable-energy-based-modern-energy-services-cuba-case-isla-de-la

GEF. (2017d). Capacity Building Programme to implement South Africa's Climate National System. Available at: www.thegef.org/project/capacity-building-programme-implement-south-africas-climate-national-system (accessed 18 December 2017).

Government of Canada, Global Affairs Canada, Deputy Minister of Foreign Affairs, Assistant Deputy Minister Public Affairs, Corporate Communications. (2017, August 22). Canada's Feminist International Assistance Policy. Available at: http://international.gc.ca/world-monde/issues_development-enjeux_developpement/priorities-priorites/policy-politique.aspx?lang=eng#3 (accessed 19 September 2017.

IDB. (2012). Institutional strengthening in support of Guyana LCDS. Available at: www.iadb.org/en/projects/project-description-title,1303.html?id=GY-G1002

IDB. (2014). Support to the implementation of National Climate Change Strategy in El Salvador. Available at: www.iadb.org/en/projects/project-description-title,1303.html?id=ES-T1219 (accessed 18 December 2017).

IDB. (2017). Institutional strengthening for the implementation of the energy reform. Available at: www.iadb.org/en/projects/project-description-title,1303.html?page=3&id=ME-T1308

National Audit Office. (2017). Managing the Official Development Assistance target – a report on progress. Available at: www.nao.org.uk/wp-content/uploads/2017/07/Managing-the-Official-development-Assistance-target-a-report-on-progress.pdf

NORAD. (2013a). Regional capacity building for sustainable natural resource management and agricultural productivity under changing climate. Available at: www.norad.no/en/front/funding/norhed/projects/regional-capacity-building-for-sustainable-natural-resource-management-and-agricultural-productivity-under-changing-climate/

NORAD. (2013b). Incorporating climate change into ecosystem approaches to fisheries and aquaculture management in Sri Lanka and Vietnam. Available at: www.norad.no/en/front/funding/norhed/projects/incorporating-climate-change-into-ecosystem-approaches-to-fisheries-and-aquaculture-management-in-bangladesh-sri-lanka-and-vietnam/

NORAD. (2013c). Sustainable natural resource management for climate change adaptation in the Himalayan Region. Available at: www.norad.no/en/front/funding/norhed/projects/sustainable-natural-resource-management-for-climate-change-adaptation-in-the-himalayan-region-a-collaborative-project-among-norway-nepal-pakistan-and-bhutan/ (accessed 18 December 2017).

NORAD. (n.d.). About Norad. Available at: www.norad.no/en/front/about-norad/ (accessed 18 December 2017).

NORAD. (n.d.) Adaptation to climate change. Available at: www.norad.no/en/front/thematic-areas/climate-change-and-environment/adaptation-to-climate-change/ (accessed 18 December 2017).

OpenAid. (n.d.) Indaba Agricultural Research Institute 2013–17. Available at: https://openaid.se/activity/SE-0-SE-6-5119006101-ZMB-31120/ (accessed 22 September 2017).

Roberts, J. T., & Weikmans, R. (2017). Postface: Fragmentation, failing trust and enduring tensions over what counts as climate finance. *International Environmental Agreements: Politics, Law and Economics*, *17*(1), 129–137.

UNEP. (2017a). Funding for UN Environment: Global Environmental Facility. Available at: http://web.unep.org/about/funding/donors-and-contributions/global-funds/global-environment-facility (accessed 18 December 2017).

UNEP. (2017b). Our work. Available at: www.undp.org/content/undp/en/home/ourwork/overview.html (accessed 18 December 2017).

UNEP. (n.d.). Ecosystem-based adaptation. Available at: http://web.unep.org/climatechange/adaptation/what-we-do/ecosystem-based-adaptation (accessed 18 December 2017).

USAID. (2014, March). Climate-resilient development: a framework for understanding and addressing climate change. Available at: www.usaid.gov/climate/climate-resilient-development-framework (accessed 19 September 2017).

VEGA. (2015). Capacity building of Cambodia's local organizations program work plan. Available at: http://pdf.usaid.gov/pdf_docs/pa00mf1k.pdf (accessed 19 September 2017).

World Bank. (2009). Bangladesh – Higher Education Quality Enhancement Project. Available at: http://projects.worldbank.org/P106216/higher-education-quality-enhancement-project?lang=en (accessed 18 December 2017).

World Bank. (2013). Kenya Water Security and Climate Resilience Project. Available at: http://projects.worldbank.org/P117635/kenya-enhancing-water-security-climate-resilience?lang=en&tab=overview

World Bank. (2014). Coastal Region Water Security and Climate Resilience Project. Available at: http://projects.worldbank.org/P145559?lang=en

World Bank. (2015). PK Sindh Irrigated Agriculture Productivity Enhancement Project. Available at: http://projects.worldbank.org/P145813?lang=en

World Bank. (2016a). Mekong Delta Integrated Climate Resilience and Sustainable Livelihoods Project. Available at: http://projects.worldbank.org/P153544/?lang=en&tab=overview

World Bank. (2016b). Integrated Disaster Risk Management and Resilience Program. Available at: http://projects.worldbank.org/P144539/?lang=en&tab=overview

World Bank. (2016c). World Bank Group Climate Change Action Plan. Available at: http://pubdocs.worldbank.org/en/677331460056382875/WBG-Climate-Change-Action-Plan-public-version.pdf (accessed 18 December 2017).

World Bank. (2017). What we do. Available at: www.worldbank.org/en/about/what-we-do (accessed 18 December 2017).

7 Universities as the central hub of capacity building

*With Shaila Mahmud,
Revocatus Twinomuhangi and
Stacy-ann Robinson*

Introduction

Although the challenge of climate change is universal, the ability to effectively mitigate climate change and cope with its effects is not. Many of the world's developing countries have limited capacities to plan and implement adequate climate policies and actions, yet it is essential that these countries build the capacity to do so. The countries described in Chapter 5 have similarities and differences in their levels of capacity and their building efforts, but those efforts mostly consisted of short-term, project-based initiatives. The measures taken thus far in the name of capacity building have been ad-hoc, uncoordinated and fragmented, with no sustainable system left in place. They have been predominantly donor-led in partnership with government organizations or NGOs, and focused mainly on involving government officials in such efforts. However, government jobs are transferable across ministries and agencies, whereas the process of capacity building must be a continuous system of production of capacities and skills over time.

The establishment of Article 11 under the Paris Agreement and the PCCB established beneath it indicate that capacity building must be based on a long-term plan. This emphasis on longevity ensures that the process remains sustainable and that the countries concerned retain ownership. The role of universities in developing countries will be central in carrying out this crucial task. So far, universities around the world have not sought to organize themselves in ways to lead major, nationally-coordinated capacity-building initiatives targeted at specific problems.

This chapter focuses on the ways universities in both the developed and the developing countries can play a role as the core of a new global capacity building initiative to address climate change. Universities have been selected from the three case countries – Bangladesh, Uganda and Jamaica – and have been reviewed to determine what kinds of capacities related to climate change education and research they currently have, and which gaps must to be filled most urgently. After a general discussion about the benefits of focusing on universities in building national capacity to address climate change, we review leading universities in Bangladesh, Uganda and Jamaica, and discuss the gaps and needs that emerge from this review. We briefly assess the relations of universities and their civil society constituencies, and then describe the launch of two new networks of universities – one in Least Developed Countries and one that is global – in 2016 and 2017. We conclude with thoughts regarding the future of these networks, for the universities, and for funding agencies and Convention staff who seek to move this effort forward.

Universities: ideally-suited hubs to build capacity

We propose that universities be a central hub for building capacity due to the unique role these institutions hold in developing countries. Not only are universities the seats of higher learning and research, they are also often the most sustainable institutions in developing countries. Many universities have outlasted the rise and fall of empires, revolutions, upheavals and diverse political regimes. They can play a critical role in building lasting, climate-related capacities around the world. Many researchers, educators, and students are already deeply involved in learning, producing and communicating climate change knowledge and skills. Identifying and publicizing effective university programmes, networking across disciplines, institutions and countries, designing new research programmes, and expanding access to educational resources and opportunities will only support the development of more resilient ecosystems and societies.

As described in Chapter 6, developed countries typically support capacity building by funding disconnected initiatives through development assistance agencies on an ad-hoc basis. They often hire consultants to conduct training sessions or give other short-term assistance, but provide little to no continuing support (Huq 2016). As attempts to implement the Paris Agreement move forward, it is crucial to consider how this pattern can be altered so that funding earmarked for capacity building constitutes not just disparate expenditures without lasting effect, but *investments* that build local capacities for decades and generations to come. Initiatives that empower universities in the developing world to effectively teach climate change could perform this function of transforming expenditures into true investments by setting up systems that will continue to build capacity for years after the funding is exhausted.

Universities can also enable students to engage with the myriad of social and economic issues related to climate change before these same students become the next generation of leaders in government, civil society and the private sector. Universities are *already* building local capacities by carrying out research and providing education on climate change. Professors often construct opportunities for students to learn skills and knowledge that will be useful in addressing climate change-related problems. Almost all forms of climate action require substantial funding, and most capacity-building efforts will be no different. However, universities have many resources that can be shared readily without seeking new and additional funding.

Almost every university department has a meaningful contribution to make regarding how best to address climate issues. However, there still exist barriers that prevent universities from effectively building capacities where they are most needed. Students from the developing world have few opportunities to attend universities in industrialized countries, and when in their own countries, often lack access to computers, databases, professional journals, and other crucial resources. Language barriers can inhibit access to necessary information in scientific and applied policy literature. Researchers tend to collaborate with others they have met in graduate school or professional conferences held in their home regions, rather than reaching out to those with different backgrounds, perspectives and resources.

South-South, South-North, and multilateral knowledge sharing among universities will benefit all participants. Low-cost, high-impact activities could include the following plans (Hoffmeister et al. 2016).

Global engagement and research collaboration

Numerous climate researchers, educators and practitioners are already working to support efforts to implement the Paris Agreement. Many other universities hope to become similarly engaged, and this aspiration should be linked with climate policy research needs, both under and outside of the Convention. Certain NGOs and negotiating groups are already playing a crucial role, requesting that their contacts at universities conduct research as new needs for information emerge under the UNFCCC (e.g. by inviting a university research group to contribute a response to a UNFCCC body's call for submissions on a particular issue) and apprising them of other areas requiring research support.

An example of an on-going collaboration between an NGO and a university group already exists between two of the organizations with which the authors of this book are affiliated – the International Centre for Climate Change and Development (ICCCAD) and Brown University's Climate and Development Lab (CDL). In early 2016, Dr Saleemul Huq, Director of ICCCAD, alerted the CDL of the Warsaw International Mechanism Executive Committee's call for submissions on financial instruments supporting loss and damage as well as the larger need for research on financing options for loss and damage (CDL & ICCCAD 2016). The subsequent project involved a workshop at the German Development Institute in Bonn, a briefing paper published by DIE, and two academic articles. The work has been included in subsequent positions taken by the WIM ExComm. Such models of collaboration could easily be emulated by other groups.

A central task of the Paris Committee on Capacity Building – perhaps working with the Climate Technology Centre and Network (CTCN)

(CTCN n.d.) – could be to help match research needs with universities and other experts in order to build capacities in targeted areas.

Problem-based collaboration

Capacity building may be best served when experts from different fields come together to work on complex problems. University researchers and educators can exchange skills and knowledge with practitioners who work with communities and know how to cater programmes to local needs.

Access to information

Universities in developing countries often have limited access to the internet, and therefore to databases, academic journals, professional development materials, and other resources. Establishing lasting relationships between universities could improve access to many types of information, thus narrowing the digital divide and strengthening the easy-access movement (Inefuku 2017).

- *Scholarly work.* Researchers and educators in developing countries often have limited access to useful information and analysis of climate-related issues, especially since access to peer-reviewed journals can be costly. Various environmental and other NGOs also produce valuable materials only accessible online. Universities in the Global North may be able to ease access for researchers and practitioners from the Global South, for example, by compiling a list of salient resources, particularly if entries are accompanied by summaries and reviews.
- *Climate-related data.* Action on climate change should be grounded in sound science. Therefore, national agencies and local communities alike need access to information about weather conditions, climate projections, environmental degradation, such as deforestation, and other issues relevant to their locale. Such information should be provided in a format useful to decision-makers and practitioners. Collaborations among researchers from the Global North and the Global South can promote discussions about how data can be most effectively leveraged in concert with local knowledge to produce effective mitigation and adaptation activities.
- *Curricular materials.* Climate-related topics are now taught in almost every university department, both in stand-alone classes and as units within other courses with other focuses. Collecting syllabi and reading lists, and making them broadly available, would make it easier for instructors unfamiliar with climate change to incorporate relevant materials into their classes. More experienced instructors can also learn valuable new approaches from their peers, and so these materials would be of use to educators throughout the world.

Distance learning

Many universities and private companies offer educational opportunities that do not require on-site participation. For example, edX offers courses addressing energy, water management, risk assessment, and many other topics relating to climate change (EdX 2017). Many of these courses are free. Compiling a list of massive open online courses (MOOCs) and other online resources relevant to climate change would be helpful to countries lacking their own climate-related courses and the capacity to develop them.

Student exchanges

Many university students in both the Global North and the Global South study abroad for a short period of time. Accordingly, universities often set up student exchange programmes that feed into valuable long-term relationships. Currently, however, tuition is prohibitive for some students wishing to study abroad. More exchanges could be accomplished without significant additional financial support if students were to work as interns rather than taking classes for credit at a foreign school (ibid.). Internships could be facilitated if students' home schools agreed to award credit for the experience, and if receiving universities provided students with mentors or supervisors. Living expenses could also be minimized if professors or other students were to provide housing for visiting students. Mentorships could lead to long-term collaborations, and such relationships could expand to include other researchers within the participating institutions. The PCCB could collect information on such programmes and suggest ways to facilitate and improve student exchanges.

Assessing capacity at universities in Bangladesh

Because of its extreme vulnerability to environmental degradation and climate change impacts, Bangladesh has been one of the front runners in devising environment and climate change-related educational programmes in a number of public and private universities. Of these, four universities based in Bangladesh's capital city, Dhaka, have been selected, two public and two private: Bangladesh University of Engineering and Technology (BUET, public), Dhaka University (the oldest public university, established in 1921), Independent University Bangladesh (IUB, private) and North South University (NSU, private). Other public universities outside of Dhaka, such as Jahangir Nagar University (close to Dhaka) and Khulna University (in southern coastal area) have distinct environment and climate change-related departments. However, due to time and resource constraints, we have reviewed only the four Dhaka-based universities here.

Environment/climate change-related teaching programmes

All of the four universities selected have different discipline-focused environmental education, each offering undergraduate and Master's degrees, either in environmental science/management or in climate change or related areas.

The Department of Water Resources Engineering has been the principal vehicle at the Bangladesh University of Engineering and Technology (BUET), the best technical university in Bangladesh, introducing climate education in its curricula. As a premiere institution of higher learning, BUET is able to attract students from countries like India, Nepal, Iran, Jordan, Malaysia, Sri Lanka, Pakistan and Palestine.

The University of Dhaka imparts knowledge on environmental and climate change-related issues through both the faculty of Earth and Environmental Sciences and the faculty of Biological Sciences. The former encompasses five departments: Geography and Environment; Geology; Oceanography; Disaster Science and Management; and Meteorology. Seven out of the ten departments under the faculty of Biological Sciences also focus on environmental aspects: Departments of Soil, Water and Environment; Botany; Zoology; Biochemistry and Molecular Biology; Microbiology; Fisheries; and Genetic Engineering and Biotechnology. The curricula of these departments focus on environmental aspects, while only the Department of Disaster Science and Management addresses climate change-related study as a separate avenue. Both BUET and the University of Dhaka also offer PhD programmes.

The Independent University of Bangladesh (IUB), one of the leading private universities in Bangladesh, has undertaken climate change and environmental management-related programmes since it was established in 1993. Currently, the School of Environmental Science and Management (SESM) offers undergraduate and graduate programmes under the departments of Environmental Science, Environmental Management and Population Environment. The school's programmes cover major aspects of environmental studies, including Water, Air and Soil. It has a specific Master's programme on Climate Change, and it is the host of the International Centre for Climate Change and Development (ICCCAD), a world leader on climate science and policy.

North South University, the first private university in Bangladesh, was established in 1992 along the lines of the American model of liberal education, and pioneered a four-year undergraduate programme in Environmental Studies. The programme, renamed Environmental Science and Management, now offers two types of degrees: a BS in Environmental Science and a BS in Environmental Management. In 2007, the department began offering a Master's programme in Resource and Environmental Management, which includes science-based courses on different aspects of environmental media as well as Environmental Economics, Law, Ethics,

GIS, etc. Every degree programme offers an interdisciplinary course on climate change, and the Department of Public Health also teaches climate change-related courses, such as Epidemiology, Climate Change and Public Health, etc.

Research related to climate change

Of the six research institutes in BUET, two are directly related to environment/climate-focused education: the Institute of Water and Flood Management (IWFM) and the BUET-Japan Institute of Disaster Prevention. The IWFM has actively undertaken capacity building and skill development projects to tackle climate change. Recently, the IWFM, in partnership with the University of Exeter and Hedley Centre, conducted a regional project called the High End Climate Impact and Extremes (HELIX) that focused on potential impacts of extreme climate events.

The Climate Change Study Cell of BUET was also established in 2007. It focuses mainly on providing education and training, and conducting research on climate-related disasters. It also facilitates research, organizes short courses, workshops and seminars, provides advisory services to the government and relevant organizations, hosts international and national conferences, and publishes journal articles, thereby developing climate change databases and creating an internet-based knowledge dissemination portal. The Hydrology and Climate Change (HCL) Research Group of BUET was set up to develop climate change scenarios for Bangladesh and study the associated impacts on agriculture and hydrology cycles. One of Dhaka University's research institutes, the Institute of Disaster Management and Vulnerability Studies, is quite active in doing research related to DRR.

IUB has a number of research centres and institutes. Among them, the International Centre for Climate Change and Development (ICCCAD) is a leader. It was established with a mission to generate and disseminate knowledge on climate change, with specific attention to adaptation. ICCCAD, led by Dr Saleemul Huq (a co-author of this book), focuses on serving the Global South. It has played a leading role in shaping and contributing to capacity-building efforts related to climate change and development through its Master's programme on Climate Change and Development, through long- and short-term courses. To bridge the existing gap and needs between the Global North and the Global South, the Master's programme includes participant universities from across the world: Imperial College, London; Institute of Development Studies, University of Sussex; University of Manchester; Columbia University; University of Melbourne; University of Toronto, Canada; and Vietnam Water Resources University. ICCCAD has been organizing the national/international Gobeshona (Research) platform since 2015, as a knowledge hub for climate change research in Bangladesh and beyond. ICCCAD also initiated the formation of the Least Developed Countries (LDCs) Universities Consortium on Climate Change (LUCCC), a South-South, long-term capacity-building programme in association with ten universities from the LDCs. The LUCCC focuses on exchanging knowledge on climate change, paying particular attention to adaptation, primarily through training and research. ICCCAD also organizes an annual Adaptation Conference in Bangladesh that focuses on youth capacity building, and a biennial International Conference on Community-based Adaptation.

North South University also has several research institutes, including the Global Health and Climate Change Institute. For the last decade, the university's departments of Environmental Science and Management and Public Health have also done some pioneering research on topics including microinsurance and climate change, dengue fever and climate change, and climate change and epidemiology, among others.

Finally, over half of the more than 1500 full-time faculty members in these four universities have PhDs from reputable universities in the West. A number of faculty members are also closely involved in national vulnerability assessments and government- and NGO-organized capacity-building initiatives, and many participate as expert members in the Bangladesh delegation to the UNFCCC negotiations.

International collaboration

Each of the four universities described in this section is involved in research collaborations with other universities across the globe.

BUET is currently involved in collaborative research, with projects such as: High-End Climate Impacts and eXtremes (HELIX), Climate Adaptation and Livelihood Protection (CALIP), and Haor Infrastructure and Livelihood Improvement Project (HILIP). HELIX is a collaborative project that began in November of 2013. It is funded by the European Union, and has 16 participating institutions led by Exeter University in the United Kingdom. HELIX conducts assessment of the climate change impacts that would result from 2 °C, 4 °C and 6 °C of warming across the globe, under a range of physical and socio-economic conditions and considering different adaptation scenarios.

IUB conducts collaborative research with universities such as Harvard and Stanford. The School of Environmental Science and Management at IUB, along with ICCCAD, also represents Bangladesh in the Center for Natural Resources Development (CNRD) network, founded by Cologne University of Applied Sciences (CUAS) in Germany. CNRD, supported by DAAD, EXCEED and the Federal Ministry of Economic Cooperation and Development, Germany, is a platform of 13 member universities from Africa, Asia, Europe, North America and South America that makes a valuable contribution to the Post 2015 Agenda and to the Sustainable Development Goals (SDGs). IUB worked with Oregon State University in the USA to set up the Universities Network for Climate Capacity (UNCC), a network of universities that calls for all countries to work together to implement Article 11 of the Paris Agreement and the Paris Committee for Capacity Building (PCCB).

The Department of Environmental Science and Management at NSU has worked with the University of Manitoba over the last decade on the Environmental Governance Capacity Building Project, which in 2007 started a Master's programme in Resource and Environmental Management. The six-year project was funded by the Canadian aid agency CIDA.

Conclusion

The four universities in Bangladesh described above either have standalone climate change degree programmes or substantial units on climate change in other environment-related degree programmes. However, while institutions of higher learning in Bangladesh have adequate faculty members, they suffer from inadequate infrastructural and logistical support. Limitations in research facilities (for example, lack of state-of-the-art labs and libraries), along with inadequate support in ICT-based resources, have restricted the ability of students and professors to carry out good quality research. Overall, the universities in Bangladesh severely lack funding to undertake research, and most of their collaborations are short-term and project-based, producing some targeted goals and results, but little more.

Climate capacity at Makarere University in Uganda

Makerere University (MAK) is the premier public university in Uganda, and one of the oldest in Africa (Makerere University 2017a). The university began in 1922 as a small technical institute, then became a University College affiliated with the University of London in 1949 and a constituent College of the University of East Africa in 1963. In 1970, the University of East Africa was split into three fully-fledged independent national universities – Makerere University in Uganda, Nairobi University in Kenya and University of Dar es Salaam in Tanzania. MAK is currently composed of ten Constituent Colleges, including the College of Agricultural and Environmental Sciences (CAES); the College of Natural Sciences; and the College of Veterinary Medicine, Animal Resource and Bio-security. MAK's 2008/09–2018/19 Strategic Plan set a vision of becoming a leading institution for academic excellence and innovations in Africa (Makerere University 2008).

Climate change teaching and training

As Uganda's premier university, MAK designs and reviews the curricula of various programmes intended to advance climate change studies. For example, in the CAES, the curriculum has been revised to integrate climate change. In addition, over 80 academic staff have been taught about climate to enable them to incorporate it into their teaching and learning. Within the CAES, departments teach undergraduate and graduate programmes related to climate change. In the Department of Environmental Management, the following programmes, each including a climate change component, are offered: BSc. Environmental Science, MSc. Environment and Natural Resources, and PhD in Environment and Natural Resources (ENR). In the Department of Geography, Geo-Informatics and Climatic Sciences, there is a Meteorology Unit that works closely with the Uganda National Meteorological Authority (UNMA), which trains meteorologists and climate scientists, and offers a BSc in Meteorology and a Post-Graduate Diploma in Meteorology. The Department has started a new programme that includes a Bachelor of Geographical Sciences, a Master of Geographical Sciences and an MSc in Disaster Risk Management that all fully integrate climate change. MAK is not yet offering a stand-alone programme in Climate Change, but an MSc in Climate Change and Sustainability is being developed by the Department of Geography, Geo-Informatics and Climatic Sciences, and is likely to be launched for the 2018/19 academic year. Currently, the only university in Uganda offering a Master's in Climate Change is Busitema University.

Under the College of Engineering, Design, Art and Technology (CEDAT), the Department of Mechanical Engineering offers an MSc in Renewable Energy with specializations in bio-energy, hydropower, solar energy, wind energy and energy efficiency in buildings (CEDAT 2017b). Makarere University Climate Change Research Institute (MUCCRI) is supporting climate change training through student boot camps and short courses and has created climate change champions across the country. In particular, MUCCRI has trained about 30 media practitioners on climate change reporting. It has also conducted short courses on various aspects of climate change (including climate change and sustainable development, climate change and agricultural development, climate change and natural resource management, and climate-resilient urban development) that have trained over 160 participants drawn from the central and local governments, academia, researchers, civil society, the private sector and development partners.

With support from the Makerere University Centre for Climate Research and Innovations (MUCCRI), MAK students have formed the Makerere University Climate Change Association (MUCCA) aiming at making MAK a sustainable and greener campus. MUCCA promotes climate mitigation through the planting of indigenous species such as prunus africana, khaya species, cordia africana, and fruit trees for afforestation, restoration and supporting of community livelihoods (Makerere University 2016).

Climate change-related research

Universities the world over are faced with the challenge of climate change and how to achieve a climate-friendly future. MAK has a Directorate of Research and Graduate Training that among other tasks coordinates and administers research, advises on research priorities, links the university to other academic and research institutions globally and mobilizes research funding (MUCCRI 2017b). Over the last decade there has been a growing interest at MAK in climate change-related research especially at the CAES, the College of Health Sciences (CHS) and the CEDAT, and much good work has been done in this respect. In 2013, the MAK Senate approved the University Research Agenda (2013–2018), that includes a multidisciplinary research agenda, and climate change is among the five key research priority areas (Makerere University 2013). The five priority research areas are: (1) health and health systems; (2) agricultural transformation, food security and livelihoods; (3) natural resource management and climate change; (4) education and education systems; (5) governance, human rights, culture and economic management; (6) science and technology; and (7) cross-cutting issues (biotechnology, knowledge transition, gender and human resource development). Although climate change is mentioned as priority area (3), it is implied in three other priority areas (1, 2, and 5), which have a bearing on climate change. In addition, the university encourages interdisciplinary and collaborative education and research programmes that accommodate environment and climate change as cross-cutting issues (Makerere University 2016).

MAK does not have a special research fund for climate change. Therefore, in order to conduct climate change research, university staff mainly respond to research calls from funding agencies. This implies that the research conducted largely responds to the donors' priorities and not those of the university. It is only the Swedish International Development Cooperation Agency (SIDA) research support that has allowed the university to conduct a needs assessment and is integrated into the MAK's priority areas. Academic staff compete for the SIDA research funding based on these priorities. Importantly, the SIDA research grants also put a lot of focus on climate change, renewable energy and poverty studies, among others. Again, through the university research agenda, MAK promotes interdisciplinary networks of experts on the environment and climate change at the local, national, regional and international levels, with the aim of collaborating on common environmental and climate-related projects both in research and education (ibid.). Academic staff members, especially in CAES, are board members of environmental NGOs that are actively involved in addressing issues of climate change. The Intra-ACP arrangement, the Regional Universities Forum for Capacity Building in Agriculture (RUFORUM) Graduate Teaching Assistantship, and other projects operate at the university.

In CEDAT, there is a Centre for Research in Energy and Energy Conservation (CREEC), which is mainly a research, consultancy and training organization. The CREEC has done a lot of work on clean cooking using improved cooking stoves and biogas, aimed at the reduction of emissions into the environment (CREEC 2017). In addition, there is a solar laboratory which tests all lamps in the market by measuring and capturing the amount of CO_2, CO and other particulate matters released by the lamps and stoves (Makerere University 2016). CEDAT (CEDAT 2017) is also privileged to host the first ever centre of excellence, the East African Centre for Renewable Energy and Energy Efficiency (EACREEE). In 2008, the Makerere University research team at CEDAT built a five-seater plug-in hybrid electric vehicle, the Kiira EV (ibid.) and in partnership with Kiira Motors built and formally launched a solar-powered bus, the Kayoola EV, in 2016. In the College of Computing and Information Sciences (COCIS) some work has been done on 'innovative application of ICTs', aiming at bridging the gap between knowledge generation and end users for climate resilience building in Africa (Makerere University 2016).

One of the most important climate change projects at Makerere University is the Rockefeller Foundation-supported project entitled 'Strengthening East African Resilience and Climate Change Adaptation Capacity through Training, Research and Policy Interventions', that was implemented from 2009 to 2014. One of the outcomes of the project was the establishment of the Makerere University Centre for Climate Research and Innovations (MUCCRI) launched in 2013. The establishment of MUCCRI was motivated by the need to strengthen climate change research, innovations and information generation and dissemination in Makerere University by providing a one-stop centre for climate change research and innovations (MUCCRI n.d.). MUCCRI operates under four major themes, namely, (1) climate science; (2) mitigation; (3) adaptation; and (4) policy, training and outreach. MUCCRI's overall goal is to develop and coordinate inter- and cross-disciplinary collaborative research and innovations related to climate change. MUCCRI is currently implementing a USAID-supported project entitled 'Education and Research to Improve Climate Change Adaptation' (ERICCA). The project is being implemented together with another USAID-supported project entitled 'Enhancing climate resilience of agricultural livelihoods', led by the International Institute of Tropical Agriculture (IITA) in Uganda and the National Agricultural Research Organization (NARO). The NARO-IITA research activities are linking with research and capacity building efforts of MUCCRI to support MSc and PhD research on climate-smart agriculture technologies and policies (MUCCRI 2017c).

With support from the EU/FAO Global Climate Change Alliance project, MUCCRI has a developed a Climate Change Adaptation Knowledge Base (CCAKB) (MUCCRI n.d.) and is also implementing a Climate Change Impact Modelling project. In 2015, with support from UN Habitat, the Department of Geography Geo-informatics and Climatic

Sciences hosted an Urban Action Innovations Lab (UAL), a knowledge-based approach to engage with issues of sustainable urban development challenges. The Lab has evolved into the UNI Climate Change Hub, with Makerere University as the anchor institution, expanding in the East African region. UAL and the Climate Change Hub are an opportunity to engage in sustainable urban development, climate change adaptation and urban planning issues faced by cities (Makerere University 2017e).

At CHS, the School of Public Health (with support from USAID) is leading the Resilient Africa Network (RAN), an initiative focusing on strengthening the resilience of communities by nurturing and scaling innovations in participating universities (RAN 2013). RAN is a partnership of 18 African universities in 13 countries (Uganda, Rwanda, Democratic Republic of Congo, Tanzania, Somalia, South Africa, Zimbabwe, Malawi, Ethiopia, Kenya, Ghana, Senegal and Mali) that applies science and technology using a resilience-based approach to programming to strengthen the resilience of African communities against natural and man-made stresses. Resilience to climate change and variability is one of the themes focused on by RAN.

With support from the EU GCCA+, Makerere University has developed a Green Strategy (Makerere University 2016) as a means to achieve a climate-friendly future. The strategy provides an appropriate platform for mainstreaming of environment and climate issues and presents a proactive approach on what Makerere University could do to reduce emissions and make the region more resilient to the climate change ahead.

Climate change-related collaborations

MAK has had numerous collaborations with the Government of Uganda Ministries, Departments and Agencies, donors, civil society and the private sector to enhance the university's capacity for climate change training, research and innovations. The most prominent partnerships on climate change are with SIDA, NORAD, FAO and USAID. The university has a strong partnership with SIDA, which supports graduate training and research activities with climate change as a cross-cutting issue. The Norwegian Agency for Development Cooperation (NORAD) has been partnering with MAK in the fields of agriculture, forestry and nature conservation (Makerere University 2017b). NORAD also supported the University to set up a Meteorology Unit in the Department of Geography, Geoinformatics & Climatic Sciences. The Unit trains meteorologists and climate scientists, and conducts research (MUCCRI 2017a). The University has a strong partnership with USAID that is supporting the RAN in the College of Health Sciences. In addition, USAID, through the ERICCA project, is enhancing the capacity of MUCCRI. Makerere University is partnering with FAO and NARO on climate-smart agricultural technologies. In addition, MUCCRI is partnering with NARO and IITA on climate change-related research in agriculture. Makerere University is partnering with FAO, USAID, GIZ and the Climate Change Department in the Ministry of Water and Environment to develop a Climate Change Knowledge Management System. Makerere University has also worked with the Climate and Development Knowledge Network (CDKN) to conduct a study on the economic assessment of the impacts of climate change in Uganda. Recently, Makerere University joined other LDC Universities from Bangladesh, Ethiopia and Tanzania to form the Least Developed Countries Consortium on Climate Change (LUCCC). The consortium was launched during the 11th Community-Based Adaptation Conference (CBA11) in Kampala on 26 June 2017.

Capacity in Jamaica and across the Caribbean: the University of the West Indies

The University of the West Indies (UWI) is the Caribbean region's premier educational institution – it is a public university serving 18 English-speaking countries in the region, including Jamaica (Hall 2014). It welcomed its first students at Mona in Jamaica in 1948 as an independent external college of the University of London (ibid,; University of the West Indies n.d.). In 1962, it gained independent degree-granting status as the UWI. The Imperial College of Tropical Agriculture in Trinidad, which it had acquired, became the site of its second physical campus (St. Augustine), and in 1963, it established the College of Arts and Science in Barbados, which later became the site of its third physical campus (Cave Hill). Its fourth but virtual campus (the Open Campus) was formalized in 2008 and currently has 50 physical locations across the region (University of the West Indies 2016a). The Open Campus focuses on continuing and professional education and offers multi-mode teaching options, including by distance learning (University of the West Indies 2016a). The University has seven priority foci (University of the West Indies 2017c). These are linked to the regional priorities identified by the Caribbean Community (CARICOM), which are critical components of regional development and which include environmental management and innovation (University of the West Indies 2017c). Apart from UWI, there are two other registered local universities in Jamaica – the University of Technology and the Northern Caribbean University (The University Council of Jamaica 2017).

Climate-related teaching, training and workshops

The Department of Physics at Mona offers two Master's degree programmes in renewable energy – a Master of Science in Renewable Energy Technology and another in Renewable Energy Management (University of the West Indies 2017e). Students in these multidisciplinary programmes are exposed to various energy sources, distribution technologies, and building and industrial processes that promote efficient energy use (University of the West Indies 2017i). They are also exposed to essential management tools. In the Renewable Energy Technology programme, for example, students are required to take courses in bioenergy, energy economics, hydro and marine power, and solar energy conversion. Energy storage and geothermal energy are among the course electives. The Department also offers related MPhil and PhD degrees. In the 2015/2016 academic year, there were 25 MPhil and PhD students in the Physics Department at Mona. Additionally, in 2004, the Mona Campus announced an intensive course in Alternative Energy that would begin in 2005 and be delivered by the Department of Physics alongside other university departments (University of the West Indies 2004). The state-owned Petroleum Corporation of Jamaica provided initial financial support and scholarships (ibid.).

The Centre for Resource Management and Environmental Studies at the Cave Hill in Barbados offers a Certificate in Climate Change in partnership with the Open Campus (Caribbean Community Climate Change Centre 2013). The programme consists of four courses: (1) Basic Concepts and Issues in Nature and Conservation; (2) Introduction to Climate Change; (3) Climate Change Policy; and (4) Sector Responses to Climate Change (University of the West Indies 2016b). The certificate was first offered in 2013 for US$500 and for US$1,000 from 2014 onwards (Caribbean Community Climate Change Centre 2013).

The UWI regularly hosts climate-related workshops including IPCC events, training and Awareness Week programmes across the three physical campuses (*The Jamaica Observer* 2016). In 2012, the Department of Physics at Mona hosted a Regional Training Workshop on Ensemble Climate Modelling. The workshop targeted government and community workers, private sector workers and students 'interested in gaining basic knowledge of climate change modelling and the use of its outputs in assessing vulnerability and adaptation to climate change' (University of the West Indies 2012a). The training manual was developed by the CSG and the Instituto de Meteorología (Cuba), and comprised five modules: (1) Climate Change Basics; (2) Modelling and Projections I: Basics; (3) Modelling and Projections II: Other Considerations; (4) Related Issues to Bear in Mind; and (5) Optional Units (e.g. Vulnerability, Introduction to Mitigation, and Regional Adaptation) (University of the West Indies and Instituto de Meteorología 2009). The workshop was funded by the European Union's Global Climate Change Alliance Project through the Caribbean Community Climate Change Centre (ibid.). The Renewable Energy Teaching and Research Laboratory at Cave Hill has mobile training toolkits that were donated by the German Federal Enterprise for International Cooperation (GIZ) (King 2015).

Climate-related research

Research conducted by the Climate Studies Group (CSG) at the Mona Campus in Jamaica focuses on increasing understanding of local, regional and global climate (University of the West Indies 2017a). Established in 1994 in the Department of Physics (now in the Faculty of Science and Technology, which was previously called the Faculty of Pure and Applied Sciences), its faculty, technical staff, consultants and postgraduate students specifically study seasonal, mean and annual climate and climate extremes along with renewable energy potential in Jamaica and the wider Caribbean region (University of the West Indies 2017d).

The CSG's research has produced outputs and outcomes that are pertinent to Caribbean environmental sustainability. Through a multi-country initiative, for example, the Group has 'produced the only dynamically downscaled climate information for the Caribbean' (University of the West Indies 2013). In 2009, the CSG won a university-wide award for the best research publication (University of the West Indies 2009). The paper, 'Features of the Caribbean Low Level Jet', helped to improve planning during the dry summer months (ibid.). The Group also undertakes community-related climate change projects and was one of the Executing Agencies for Jamaica's Community-Based Adaptation Programme, which was implemented by the United Nations Development Programme and funded by the Global Environment Facility's Strategic Priority on Adaptation to the amount of US$245,965 (United Nations Development Programme 2017a, 2017b). UWI's sub-project, '"Tell It": Disseminating Caribbean Climate Change Science and Stories', '"translate[d]" three studies/pilot projects that have been undertaken in Jamaica into "scientific" language in order to facilitate a peer review of the work and ultimately inclusion in the Fifth Assessment Report of the Intergovernmental Panel on Climate Change [IPCC]', which was published in 2014 (United Nations Development Programme 2017b). In addition to this scientific article, the sub-project developed and disseminated a climate- and vulnerability-related non-technical report, summary for policy-makers, and short videos (ibid.).

The current Director of the CSG, who was appointed to the post in 2007, is also Head of the Physics Department and Deputy Dean of the Faculty of Science and Technology (University of the West Indies 2013, 2017i, 2017j). The Director, a veteran contributor to the IPCC process, also has a strong grant-getting track record, including leading a successful US$10.4 million grant application on behalf of the Mona Campus in 2015 (University of the West Indies 2017j). During Climate Change Awareness Week 2016, the UWI launched a supercomputer, SPARKS, valued at US$742,376 (Neufville 2017). SPARKS is expected to help the CSG generate 'faster simulations at higher resolutions', which will provide 'more accurate and credible data, and information that will improve climate projections in the short, medium and long term'; in addition, 'climate research and downscaling methods will no longer be limited by the available hardware and software' (University of the West Indies 2016c). Before SPARKS, CSG researchers only had the capacity to conduct single data runs at a time, which could take up to six months due to limited data storage (Neufville 2017). These same data runs can now be completed in around two days as the supercomputer has the capacity of over 5,000 compact discs (University of the West Indies 2016c). The acquisition of SPARKS was a partnership between Fujitsu® and Dell® through the Inter-American Development Bank's (IDB) implementation of its Pilot Programme for Climate Resilience programme of work (ibid.). The University's Office for Research and Innovation managed the project (ibid.).

The Department of Physics at Mona has a Solid State Electronics Research Laboratory, which focuses on solar energy and photovoltaic studies (University of the West Indies 2017b). In an attempt to develop and utilize 'alternate energy sources through photovoltaic cells', the Laboratory has recently 'undertaken a feasibility study for the design, development, installation and use of large-scale photovoltaic energy' (University of the West Indies 2017l: 28). The study was financially supported in part by the Mona Institute of Applied Sciences through the Canadian High Commission's Green Fund Project (ibid.). The Laboratory also focuses on wind energy studies (University of the West Indies 2017b) – in the past, it estimated Jamaica's wind resources (University of the West Indies 2017l). Through a grant from the Environmental Foundation of Jamaica, the Laboratory has completed a first-phase assessment of wind-generated electricity at a rural site and has plans to study the feasibility of establishing wind farms in other areas (ibid.). Additionally, the Faculty of Science and Technology at Cave Hill opened a Renewable Energy Teaching and Research Laboratory in 2013 (Madden 2013; University of the West Indies 2017l). The Laboratory aims to link UWI researchers with private renewable energy companies to test renewable energy technologies and to make them suitable for the local environment (Madden 2013). Researchers at the St. Augustine Campus are also undertaking research in geothermal energy, solar photovoltaic, solar thermal energy and wind energy (University of the West Indies 2017l).

UWI has a Disaster Risk Reduction Centre, which is based at Mona. The Centre was launched in 2006 and is 'an Institutional Focal Point for harnessing the expertise of the University to develop and implement training, research, advisory and outreach services' (University of the West Indies 2006a). Students studying for a Master's degree in the Faculty of Science and Technology's Natural Resource and Environmental Management programme can specialize in one of the three streams offered by the Centre: (1) Disaster Risk Management; (2) Integrated Urban and Rural Environmental Management; and (3) Marine and Terrestrial Ecosystems (University of the West Indies 2012b). The Centre has over 20 adjunct faculty members who carry out research in a number of areas, including coastal dynamics and management, disease control, economic planning, water and air quality, and psychosocial and emergency care (University of the West Indies 2017k).

Climate-related collaborations

UWI has collaborative links with around 160 universities globally (University of the West Indies 2017c). In 2016, the University launched a partnership with the State University of New York (SUNY) – a Center for Leadership and Sustainable Development at the SUNY Global Center in New York, which aims to facilitate joint research relevant to governance and leadership (University of the West Indies 2017l). There are many research projects currently underway, including one on climate change and another on coral reef health (University of the West Indies 2017). There are plans for the Centre to offer joint degree programmes and professional training in the future (University of the West Indies 2017).

The University has a Pro Vice-Chancellor of Academic-Industry Partnership and Planning who has a strong business and entrepreneurship background and who is responsible for forging and maintaining links with industries (University of the West Indies 2017l). Funding from the IDB and the European Union has boosted the University's renewable energy programme, including supporting different technologies, management and entrepreneurship in the sector (ibid.). The four-year project is being implemented across the three physical campuses (ibid.).

Gaps and needs as evident from the review

The review above of six universities – four from Bangladesh, which include two public and two private universities, and one public university each from Uganda and Jamaica, covered mainly three areas: environment/climate change-related programmes offered, related research projects/works done/being done and international collaborations.

In line with the framework of multiple roles of universities laid out in this chapter, the universities reviewed are involved with many of those functions. Some of these universities offer stand-alone degree programmes not just on environmental studies/science, but also on climate change, both at undergraduate and graduate levels. The MAK and the UWI also offer degrees in Renewable Energy, directly related to mitigation of climate change. The MAK and UWI also have distance learning programmes, to reach out to a wider audience of studentship. However, based on experiences in Bangladesh, it can be observed that student enrolment in those disciplines is not encouraging, either in environment or climate change degree programmes, particularly in private universities. Enrolment in public universities is somewhat different, as in those universities' tuition fees are just nominal amounts. Discussion with faculty members, students and experts reveal that students are worried more about finding jobs after graduation, which is still not very promising and not expanding much. Still, those who do well are absorbed by appropriate national or international agencies.

Different institutes and centres dedicated to climate research are also involved in many different partnership projects with regional and western universities funded by bilateral or multilateral agencies. The research focus obviously reflects the vulnerabilities of the three case countries and their many different solutions. The funding sources to some extent reflect colonial ties of these three countries, but they are also funded by organizations like the UNDP or the GEF.

However, the universities we reviewed are missing some elements of the ideal roles laid out above, such as regular student exchanges from South to North, and faculty exchanges also from South to North. This is very important in view of establishing a genuine partnership which will gradually germinate ownership of the capacity-building projects. A one-way track of students going to the South to take courses or doing fieldwork for their theses writing again is indicative of a transfer of knowledge from the Global North to the Global South. More exchanges of students and faculty members from the Global South to the Global North will give them free access to the latest peer-reviewed knowledge, thereby building their capacities. Also no collaboration was evident from the review of any western partner university offering free access to its southern counterparts to ICT resources, particularly access to peer-reviewed literature. This is again extremely important, otherwise the southern partners will remain less able to do research and publish in peer-reviewed journals, challenging the dominant theses and assumptions dominated by the Global North.

These weaknesses in the southern universities are reinforced by lack of national-level funding for research and building of research labs or climate observation systems or weather stations. Donors also do not often fund building of capacities at universities on a sustainable level. Finally, there is a problem in some cases of adequate language efficiency required to do advanced level research or present publications in prestigious journals.

Civil society engagement in capacity building

The UNFCCC recognizes nine civil society constituencies, including business and industry NGOs (BINGOs), environmental NGOs (ENGOs), farmers, local and municipal governments, indigenous peoples' organizations (IPOs), trade unions (TUNGOs), research and independent NGOs (RINGOs), women and gender, and youth NGOs (YOUNGOs). Organizations select a constituency to join when they apply to become UNFCCC observers. Universities, think-tanks, and consulting groups tend to choose the RINGO constituency, which welcomes anyone conducting research on any aspect of climate change. Climate action and effective capacity building will require collaboration across geographic, political, sectoral, disciplinary, and other boundaries, and will require new perspectives on programmes and policies. As researchers, educators, and practitioners, RINGO members have an ongoing interest and role in capacity building. They also have colleagues at their home institutions working on virtually every aspect of climate change with diverse approaches, who could be recruited to participate in capacity-building efforts. The RINGOs have also stressed the importance of collaboration across constituencies and are seeking ways to work more effectively with other actors outside of the negotiations process. The PCCB could benefit from the involvement of civil society constituencies in numerous ways.

First, it could invite constituency members to participate as working members of the PCCB. Second, it could co-sponsor various activities with one or more constituencies to ensure that different types of expertise are represented and a wide variety of topics addressed. Third, it could consult with the constituencies to identify experts and other resources appropriate for specific tasks. Finally, the PCCB could request that the constituencies communicate with their membership about the importance of and need for capacity building, and to solicit materials, suggestions, and other support.

Launching of the LUCCC and UNCC

During the COP22 in Marrakech, the Independent University of Bangladesh (IUB) and the University of Cadi Ayyad in Morocco held a side event to discuss the role of universities in implementing Article 11 of the Paris Agreement. The Research and Independent NGOs (RINGOs) group, along with IUB and the UNFCCC Secretariat, held another side event on the same topic. Following overwhelming support from students and faculty members of the universities from the LDCs, the International Centre for Climate Change and Development (ICCCAD) at IUB and the Makarere University Climate Change Research Innovation (MUCCRI) took the initiative to set up the LDC Universities Consortium on Climate Change (LUCCC), a South-South Consortium of ten universities to enhance knowledge on climate change through capacity building, with a focus on adaptation measures, such as education and research.

The objectives of the LUCCC are as follows:

- To foster a South-South collaborative network to enhance research capacity and proficiency in climate change. The Consortium is initiated by the LDCs and will only include the LDC universities.
- To network and develop the capacity of the South-South Consortium of universities to develop common research projects and implement teaching and training programmes in different climate change aspects. The focus of the Consortium will be on climate change adaptation, especially community-based adaptation.

- To work with the vulnerable communities for the most vulnerable communities. This will be a two-way collaborative capacity-building programme which will offer help to others and will also seek help from others in order to build capacity of their own.
- To provide capacity-building support to LDC universities to serve as a repository. The core ten universities from the ten LDCs will in turn engage with other universities in their respective sub-regions in Africa and Asia. Thus, by using a hub-and-spoke approach, all 48 LDCs will be covered over the next few years. The objectives of the LUCCC are to implement Article 11 of the Paris Agreement on Climate Change which focuses on the need to develop in-country capacity-building systems in each and every country to enable them all to tackle climate change by 2030. The LUCCC initiative aims to enable exchange of faculty, researchers and students across the LDCs to start with, but also with other universities in non-LDC developing countries as well as developed countries over time. The student-to-student exchange of knowledge and activities has already been started by the students of Makerere University, who held a separate Youth Conference on Climate Change in parallel with CBA11 in Kampala with cooperation and participation from youth from Bangladesh and other countries. Each of the LUCCC universities has agreed to hold an annual national youth conference in each of their countries where they will share knowledge on innovative practices to tackle climate change from the youth of other LDCs. Another activity will be a LUCCC Fellowship Programme where young researchers and faculty from each LUCCC partner will be able to visit and collaborate with other partners in the consortium.

Another new university network, the University Network on Climate Change (UNCC), connects the universities of both the Global South and the Global North. The secretarial services for this network are provided by the ICCCAD at IUB in Dhaka and Oregon State University in the USA.

Conclusion

As climate change intensifies, relevant skills and knowledge should be shared as widely as possible to enhance the ability of vulnerable nations to mitigate and respond to impacts of global warming. Universities have a critical role to play in this capacity building to tackle long-term climate change. Every country in the world – especially those that are poor and vulnerable – is home to universities that have not attained their desired potential to reach students and train them in climate-related issues. By contrast, a number of universities, in developed and developing nations alike, are already teaching effective courses on various aspects of the climate problem. Investing in building the capacity of universities to teach students about climate change and train them to tackle associated social, economic and political issues will be a cost-effective means of embedding long-term capacity-building systems in every country in the world. Universities, for the reasons explicated above, stand out as front runners to serve as the hub of such efforts.

At this stage, a few initiatives can be collectively thought of under the heading actions to comply with Article 11 of the Paris Agreement:

- A list of possible funding sources, both bilateral and multilateral, to support universities and students in the developing world, as well as to promote sharing of expertise and resources between the Global North and the Global South.
- To begin with, LUCCC partners should come out with an operational strategy encompassing a phased approach to building capacities first among the partner universities. Some funding should be mobilized on an immediate basis to jump-start the LUCCC partnerships across the partner universities as well as with universities of the Global North. This will showcase the potential of the LUCCC in near terms, which can be replicated beyond the LUCCC partners, between the wider UNCC partners.
- A strategy for identifying, collecting, and making available materials such as syllabi that could be useful to anyone wishing to teach climate-related issues.
- Devising strategies to expand and make the best use of the UNFCCC and UNCC networks to promote collaboration in education research and innovation.

References

Caribbean Community Climate Change Centre. (2013). CERMES/UWI Open Campus announce Climate Change Certificate Programme. Available at: https://caribbeanclimateblog.com/2013/06/05/cermesuwi-open-campus-annouce-climate-change-certificate-programme/ (accessed 15 July 2017).

Centre for Research in Energy and Energy Conservation (CREEC). (2017). Our mission. Available at: http://creec.or.ug/

Climate Development Lab (CDL) & International Centre for Climate Change and Development (ICCCAD). (2016). Financing options for loss and damage: a review and roadmap. Available at: https://unfccc.int/files/adaptation/groups_committees/loss_and_damage_executive_committee/application/pdf/browncdl-icccadfinancinglossanddamagepaperdraft.pdf

Climate Technology Centre and Network (CTCN). (n.d.) Connecting countries to climate technology solutions. Available at: www.ctc-n.org/ (accessed 16 June 2016).

College of Engineering, Design, Art and Technology (CEDAT), Makerere University. (2017a). Center for Research in Transportation Technologies. Available at: https://cedat.mak.ac.ug/research/crtt

College of Engineering, Design, Art and Technology (CEDAT), Makerere University. (2017b). Home. Available at: http://cedat.mak.ac.ug/
College of Engineering, Design, Art and Technology (CEDAT), Makerere University. (2017c). Master of Science in Renewable Energy. Available at: https://cedat.mak.ac.ug/graduate-programmes/master-of-science-in-renewable-energy
EdX. (2017). Home. Available at: www.edx.org/ (accessed 16 October 2017).
Hall, D. (2014). *The University of the West Indies: A quinquagenary calendar 1948–1998*. Mona: University of the West Indies Press.
Hoffmeister, V., Averill, M., & Huq, S. (2016 July). The role of universities in capacity building under the Paris Agreement. The International Centre for Climate Change and Development (ICCCAD) at the Independent University, Bangladesh (IUB) and Brown University's Climate and Development Lab (CDL).
Huq, S. (2016) Why universities, not consultants, should benefit from climate funds. *Climate Home*. 17 May 2016. Web. 18 May 2016.
Independent University of Bangladesh (n.d.). About. Available at: www.iub.edu.bd/AboutIUB/ataglance (accessed 29 July 2017).
Inefuku, H. W. (2017). Globalization, open access and the democratization of knowledge, *Educauset Review,* July/August.
Jamaica Observer. (2016). GOJ, UWI host Inter-Governmental Panel on Climate Change on Caribbean visit. Available at: www.jamaicaobserver.com/environment/GOJ-UWI-host-Inter-Governmental-Panel-on-Climate-Change-on-Caribbean-visit_82071 (accessed 29 July 2017).
King, L. (2015). Boost for renewable energy training. Nation Publishing Company Limited, Available at: www.nationnews.com/nationnews/news/68130/boost-renewable-energy-training (accessed 31 July 2017).
Madden, M. (2013). UWI, Cave Hill opens renewable energy teaching and research laboratory. *Barbados Today*, Available at: www.barbadostoday.bb/2013/12/17/81468/ (accessed 31 July 2017).
Makerere University. (2008). Makerere University Strategic Plan 2008/09–2018/19. Available at: https://pdd.mak.ac.ug/images/PDD_Docs/Makerere-University-Strategic-Plan2008-09-2018-19.pdf
Makerere University. (2013). The University Research Agenda: 2013–2018. Directorate of Research and Graduate Training. Available at: https://policies.mak.ac.ug/sites/default/files/policies/Research%20Agenda%202013-2018%20-%20Approved%20by%20Senate%20Dec%209%2C%202013.pdf
Makerere University. (2015). Annual Report 2015. Available at: https://pdd.mak.ac.ug/docs/annual-reports/Makerere_University_Annual_report_2015.pdf
Makerere University. (2016). A green strategy for Makerere University. Department of Environmental Management, College of Agricultural and Environmental Management with support from the Global Climate Change Alliance + Climate Support Facility.
Makerere University. (2017a). Meteorology Unit. Available at: http://geography.mak.ac.ug/centers-units/meteorology-unit/
Makerere University. (2017b). Home. Available at: www.mak.ac.ug/
Makerere University. (2017c). Research projects. Available at: www.mak.ac.ug/research/research-projects
Makerere University. (2017d). Urban Action Innovations Lab. Available at: http://geography.mak.ac.ug/centers-units/urban-action-innovations-lab/

Makerere University. (2017e). Research directorate. Available at: www.mak.ac.ug/research/research-directorate

MUCCRI (n.d.a) Home. Available at: www.muccri.mak.ac.ug/www.gcca.eu/sites/default/files/ACP/workorders/wo51_-_uganda_-_green_strategy_for_makerere_university.pdf

MUCCRI. (n.d.b). Climate Change Adaptation Knowledge Base. Available at: http://muccri.mak.ac.ug/content/climate-change-adaptation-knowledge-base (accessed 16 October 2017).

MUCCRI. (n.d.c). On-going research. Available at: www.muccri.mak.ac.ug/content/going-research (accessed 16 October 2017).

Neufville, Z. (2017). Super computer plugs gap in Caribbean climate research. *Pride News Magazine*, Available at: http://pridenews.ca/2017/03/13/super-computer-plugs-gap-caribbean-climate-research/ (accessed 1 August 2017).

Resilient Africa Network (RAN). (2013). Available at: www.ranlab.org/ (accessed 16 October 2017).

University Council of Jamaica. (2017). Registered institutions and training units as at July 2017. Kingston: The University Council of Jamaica.

United Nations Development Programme. (2017a). Community-Based Adaptation: Jamaica. Available at: www.adaptation-undp.org/projects/spa-community-based-adaptation-jamaica (accessed 16 July 2017).

United Nations Development Programme. (2017b). 'Tell it': Disseminating Caribbean climate change science and stories. Available at: www.adaptation-undp.org/projects/spa-%E2%80%9Ctell-it%E2%80%9D-disseminating-caribbean-climate-change-science-and-stories (accessed 16 July 2017).

University of the West Indies. (2004). UWI & PCJ collaborate to deliver alternative energy course. Available at: www.mona.uwi.edu/marcom/newsroom/entry/2874 (accessed 31 July 2017).

University of the West Indies. (2006a). UWI and UNDP sign joint proposal for UWI Disaster Risk Reduction Centre. Available at: www.mona.uwi.edu/marcom/uwinotebook/entry/1776SW (accessed 31 July 2017).

University of the West Indies. (2006b). UWI Disaster Risk Reduction Centre: About us. Available at: www.uwi.edu/drrc/aboutus.aspx (accessed 31 July 2017).

University of the West Indies. (2007). Professor Anthony Chen shares in prestigious Nobel Peace Prize. Available at: www.mona.uwi.edu/principal/node/47 (accessed 31 July 2017).

University of the West Indies. (2009). Climate Studies Group Mona: The best research publication. Available at: http://isis.uwimona.edu.jm/research/awards/2009/climate-studies-group-mona.php

University of the West Indies. (2012a). Upcoming events: Climate change workshop. Available at: www.mona.uwi.edu/physics/ccworkshop (accessed 30 July 2017).

University of the West Indies. (2012b). UWI Disaster Risk Reduction Centre: Graduate programmes. Available at: www.uwi.edu/drrc/gradprogrammes.aspx (accessed 31 July 2017).

University of the West Indies. (2013). UWI Mona's Prof. Michael Taylor awarded Silver Musgrave Medal for work in climate change. Available at: www.mona.uwi.edu/marcom/newsroom/entry/5294 (accessed 2 August 2017).

University of the West Indies. (2015). *Higher education and statistical review: Issues and trends in higher education, 2013*. St. Augustine: University of the West Indies.

University of the West Indies. (2016a). About the UWI Open Campus. Available at: www.open.uwi.edu/about (accessed 28 July 2017).

University of the West Indies. (2016b). Certificate in Climate Change. University of the West Indies, accessed July 28, 2017. Available at: www.open.uwi.edu/programmes/certificate-climate-change

University of the West Indies. (2016c). SPARKS launches UWI as a 'Big Deal' in climate data computing. Available at: www.mona.uwi.edu/publications/monanews/article3.html (accessed 1 August 2017).

University of the West Indies. (2017a). About CSGM. Available at: www.mona.uwi.edu/physics/csgm/about (accessed 31 July 2017).

University of the West Indies. (2017b). Alternative energy: Research. Available at: www.mona.uwi.edu/physics/csgm/about (accessed 31 July 2017).

University of the West Indies. (2017c). *Annual Report 2015–2016 – UWI St. Augustine*. St. Augustine: University of the West Indies.

University of the West Indies. (2017d). Climate Studies Group Mona. Available at: www.mona.uwi.edu/physics/csgm/home (accessed 31 July 2017).

University of the West Indies. (2017e). Department of Physics: MSc. Programme. Available at: www.mona.uwi.edu/physics/msc-programme (accessed 31 July 2017).

University of the West Indies. (2017f). Doctor of Philosophy – Physics. Available at: www.mona.uwi.edu/physics/phd (accessed 31 July 2017).

University of the West Indies. (2017g). Master of Philosophy – Physics. Available at: www.mona.uwi.edu/physics/master-philosophy-physics (accessed 31 July 2017).

University of the West Indies. (2017h). Master of Science – Renewable Energy Technology. Available at: www.mona.uwi.edu/physics/master-science-renewable-energy-technology (accessed 31 July 2017).

University of the West Indies. (2017i). Physics: Course principals.. Available at: www.mona.uwi.edu/physics/ccworkshop/course-principals (accessed 2 August 2017).

University of the West Indies. (2017j). *Report: Faculty of Science and Technology for the year ending July 31, 2016*. Mona: University of the West Indies.

University of the West Indies. (2017k). UWI Disaster Risk Reduction Centre: Staff. Available at: www.uwi.edu/drrc/staff.aspx (accessed 31 July 2017).

University of the West Indies. (2017l). *Vice-Chancellor's Report to the University Council 2015/2016*. Mona: University of the West Indies.

University of the West Indies. (n.d.) History. www.mona.uwi.edu/content/history. (accessed 28 July 2017).

University of the West Indies, and Instituto de Meteorología. (2009). *Training manual: climate modelling. The science of climate change and projections*. Available at: www.mona.uwi.edu/physics/sites/default/files/physics/uploads/Training%20Manual.pdf (accessed 30 July 2017).

8 Capacity building and transparency under Paris

With Stacy-ann Robinson

Introduction

Who is taking action to address climate change? Who is providing adequate support to developing countries to help them cope with climate impacts and green their economies? Without transparency systems, we have no way of knowing whether countries are making progress towards these ends, improving upon their efforts and reaching the levels of action needed. But in many parts of the world, the ability to reliably collect, compile, and report information on what's going on with climate change action and support simply does not exist.

The Paris Agreement lays out enhanced reporting and transparency requirements in Article 13, which are crucial to ensure that effective mitigation and adaptation actions are taken under the Agreement (Van Asselt et al. 2016, 2017). The Paris requirements potentially represent a significant step forward in systematizing what has, to this point, been somewhat patchy reporting on climate action. As Van Asselt et al. (2017) put it:

> The regular provision of this information, and a subsequent review by experts to ensure that information is reliable, have become one of the backbones of international climate agreements. By making clear what Parties are doing to implement their commitments under international agreements like these, transparency helps to build trust and confidence. Transparency can indicate whether the level of collective efforts undertaken by countries is adequate to address climate change, by shining a light on what they do individually.

The extent of progress on transparency of climate action that will occur under the Paris Agreement depends upon work going on at the time of this writing to develop 'modalities of reporting' for action and support. This work is occurring under the Subsidiary Body for Scientific and Technological Advice (SBSTA), a subsidiary body of the UNFCCC.

The Paris Agreement put in place three levels of reporting requirements, including sets for developed and developing nations, and a third much more flexible set for the world's most vulnerable and poor countries. In further recognition of the fact that some nations do not presently have the capacity to fulfil the Agreement's enhanced requirements for transparency, the Paris decision text established the Capacity Building Initiative for Transparency (CBIT) to address capacity-related needs arising under Article 13. The Global Environment Facility, housed at the World Bank, was tasked with supporting the establishment and operation of the CBIT.

The June 2016 meeting of the Global Environment Facility Council approved the establishment and programmatic directions of the CBIT Trust Fund, which was initially capitalized with about US$53 million (GEF 2016b). The expectation was that the CBIT fund would go towards identifying key problems in developing nations' reporting systems and pursuing pathways to improve upon them. The GEF was tasked with prioritizing projects in countries most in need of capacity building for transparency-related actions, in particular, the LDCs and SIDS.

This chapter begins by describing how transparency requirements and the Capacity Building Initiative on Transparency were laid out in the Paris Agreement's provisions and decisions. We then go on to update what has gone on since Paris, as the pace of developments in this area has been fairly rapid in its wake. We describe the tools, modalities, procedures and guidelines for transparency as they exist in the post-Paris world, including Biennial Reviews, Biennial Update Reports, International Assessment and Review, International Consultation and Analysis, and so on. We then build upon the case studies explored in previous chapters, briefly reviewing areas of strength and weakness in existing systems of reporting climate action taken and support received in Bangladesh, Uganda, and Jamaica. We conclude the chapter by discussing possible future directions of transparency efforts under the UNFCCC, and how transparency-related capacity building work will be a crucial determinant of whether the new Paris requirements succeed or fail.

Capacity building for transparency under the Paris Agreement

This section contains a review of the Paris Agreement's specific provisions and decisions on transparency-related capacity building. While sometimes tedious, such a review provides an essential basis for discussion of post-Paris developments on capacity building for transparency, measurement and reporting capacities in case study countries, and future directions of relevant efforts under the UNFCCC.

Article 13: Enhanced transparency framework for action and support

Article 13 of the Paris Agreement establishes 'an enhanced transparency framework for action and support', with 'built-in flexibility' to account for the Parties' different capacities for ensuring transparency and to recognize the 'special circumstances of the least developed countries and small island developing States', Parties agreed that this framework would serve to 'build mutual trust and confidence' and promote effective implementation of the Agreement by enhancing and building on existing transparency arrangements under the Convention, including 'national communications, biennial reports and biennial update reports, international assessment and review and international consultation and analysis'. Article 13 states that the purpose of the framework for enhanced transparency of action is to 'provide a clear understanding of climate change action' under the Convention, including adaptation actions and progress towards achieving the goals laid out in the Parties' nationally determined contributions, to inform the global stocktake the Agreement scheduled for 2023. The purpose of the framework for transparency of support is to 'provide clarity on support provided and received' by Parties 'in the context of climate change actions' under Agreement articles pertaining to mitigation, adaptation, finance, technology development and transfer and capacity building.

Article 13: Required reporting

Article 13 stipulates that each Party *shall* regularly provide inventory reports of national greenhouse gas emissions and removals by sinks and information necessary to track progress towards its NDC. It states also that each Party *should* provide 'information related to climate change impacts and adaptation'. Finally, 'Developed country Parties shall, and other Parties that provide support should, provide information on financial, technology transfer and capacity-building support provided to developing country Parties', while 'Developing country Parties should provide information' on the same types of support 'needed and received'.

Article 13: Technical expert review

Information submitted on emissions, progress towards NDC goals, and support provided shall undergo a technical expert review is to 'identify areas of improvement' for each Party and review the consistency of submitted information with the modalities, procedures and guidelines for transparency of action and support agreed at the first meeting of the Parties to the Paris Agreement (CMA1), taking flexibility allowances according to Parties' capacity limitations into account. The review 'shall pay particular attention to the respective national capabilities and circumstances of developing country Parties' and is to include assistance in identifying capacity-building needs for those 'developing country Parties that need it in the light of their capacities'.

Decision text: Capacity Building Initiative for Transparency (P. 12)

Article 13 states that 'support shall be provided to developing countries for the implementation of this Article' and that 'support shall also be provided for the building of transparency-related capacity of developing country Parties on a continuous basis'. Accordingly, the Paris Agreement's associated decision text establishes a Capacity Building Initiative for Transparency (CBIT) to 'support developing country Parties, upon request, in meeting enhanced transparency requirements as defined in Article 13' (para. 84).

The decision text states that the CBIT will aim to 'strengthen national institutions for transparency-related activities in line with national priorities', 'provide relevant tools, training and assistance for meeting the provisions stipulated in Article 13 of the Agreement', and 'assist in the improvement of transparency over time' (para. 85). It requests that the Global Environment Facility (GEF) support the establishment and operation of the CBIT as a 'priority reporting-related need', including 'through voluntary contributions' in future GEF replenishment cycles. The decision text stipulates also that 'developing country Parties shall be provided flexibility' in the implementation of Article 13, including in the scope, frequency, and level of detail of reporting and in the scope of review. However, all Parties except for LDCs and SIDS shall submit the information required by Article 13 no less frequently than on a biennial basis.

Next, the decision text requests the Ad hoc Working Group on the Paris Agreement to 'develop recommendations for modalities, procedures and guidelines' for the transparency of action and support detailed in Article 13, to be completed no later than 2018 and considered at COP24. The text states that this process should take into account 'the importance of facilitating improved reporting and transparency over time', as well as the needs to:

- provide flexibility to those developing country Parties that need it;
- promote transparency, accuracy, completeness, consistency and comparability;
- avoid duplication as well as undue burden on the Parties and the secretariat;
- ensure that the Parties maintain at least the frequency and quality of reporting in accordance with their respective obligations under the Convention;
- ensure that double counting is avoided;
- ensure environmental integrity.

In its development of modalities, procedures and guidelines, the Ad hoc Working Group on the Paris Agreement should also include consideration of:

- the types of flexibility available to those developing country Parties that need it;

- consistency between the methodology for reporting communicated in NDCs and the actual methodology used to report on progress made towards NDCs;
- reporting on adaptation action and planning, 'with a view to collectively exchanging information and sharing lessons learned';
- 'support provided, enhancing delivery of support for both adaptation and mitigation', 'enhancing the reporting by developing country Parties on support received', and taking into account issues considered by SBSTA on methodologies for reporting on financial information;
- information in the biennial assessments and other reports of the standing Committee on finance and other relevant bodies under the Convention;
- information on the social and economic impact of response measures.

The text then requests that the Ad hoc Working Group 'enhance the transparency of support provided in accordance with Article 9' (the finance article) of the Paris Agreement. Finally, the text decides that the Working Group's modalities, procedures and guidelines shall be applied upon entry into force of the Paris Agreement and supersede the measurement, reporting and verification system previously operational under the Convention.

Post-Paris developments on transparency-related capacity

Because the effectiveness of the Paris Agreement's transparency framework will play a large part in determining the Agreement's overall success, tenets of transparency have been debated centrally at post-Paris UNFCCC meetings, including COP22, held in Marrakech, Morocco, in November 2016, and the 44th and 46th meetings of the Subsidiary Bodies, convened in Bonn, Germany, in May 2016 and 2017. Aside from developments on the CBIT, the aspect of these transparency discussions that has proven most relevant to capacity building in the post-Paris climate regime is the interpretation of the flexibility allowance in the Paris transparency framework. Operationalizing this flexibility will require Parties to acknowledge differences in their existing capacities for transparency, consider how the Convention's principle of differentiation applies to transparency, and grapple with the dilemma that more uniform transparency requirements will necessitate more extensive capacity building for transparency.

At the 44th meeting of the Subsidiary Bodies (SB44), held in May 2016, Parties met in a contact group on transparency to discuss modalities, procedures and guidelines for the Paris Agreement's frameworks for transparency of action and support. Central to this contact group discussion were questions of how to operationalize flexibility, with developed and developing countries alike acknowledging the need to address the issue by calling for consideration of existing differences in capacities for transparency, especially among SIDS and LDCs (IISD, May 2016).

More specific statements on how to operationalize flexibility, however, suggested a greater developed-developing country divide. India, for the Like-Minded Developing Countries, called for systematically integrating flexibility into the modalities, procedures and guidelines of the Paris transparency framework in order to operationalize differentiation and the principle of Common but Differentiated Responsibilities and Respective Capacities (CBDR&RC). Saudi Arabia advocated the systematic application of flexibility on top of the flexibility allowances already functioning under the Convention. The United States stated that flexibility should only be discussed in the 'context of common procedures'. Meanwhile, Norway and the EU spoke up against 'categorical application of flexibility' (IISD, May 2016). At the conclusion of the session, there was 'general recognition' among Parties that the Paris transparency framework should build on the Convention's existing measurement, reporting and verification (MRV) framework (made up of international consultation and analysis and international assessment and review processes), but Parties had yet to make progress on the CBIT or hammer out any details of operationalizing flexibility (IISD, May 2016).

At COP22, held in Marrakech, Morocco, in November 2016, the CBIT was declared 'open for business' by the CEO of the GEF, Naoko Ishii (GEF, 2016a). Parties were at odds in negotiations on text providing guidance on funding the CBIT to the GEF, with controversy surrounding the possibility of adding after-text stating that 'support for the Capacity Building Initiative for Transparency (CBIT) will be included in the seventh replenishment', the insertion 'as additional resources to be set aside' (IISD, Nov. 2016). The final guidance to the GEF submitted by Parties to COP22 read: 'Parties welcomed the GEF Council decisions to establish the CBIT Trust Fund, to approve the CBIT programming directions, and to ensure support for the CBIT will be included in the seventh replenishment, to complement existing support under the GEF (GEF 2017c). In a COP decision, the Parties also requested the GEF continue providing information on the CBIT's establishment and operation, and enhancing capacity development in Least Developed Countries for the submission of project proposals to the CBIT, with a focus on identifying possible funding sources and enhancing long-term domestic institutional capacities (Decision FCCC/CP/2016/L.7).

During COP22, the Parties also continued discussions on modalities, procedures and guidelines for the Paris Agreement's transparency frameworks for action and support, including the operationalization of the flexibility allowance embedded in the Paris text. On the issue of flexibility, a developed-developing country divide again emerged, with Canada and New Zealand calling for consideration of flexibility in the context of each element of the new transparency framework, while China, with Saudi Arabia for the Arab Group and the Philippines, stated that the 'bifurcated structure' of the Convention's existing monitoring, reporting and verification system should be adopted as a starting point for operationalizing flexibility, such that differentiation is systematically and consistently incorporated into the structure of the new transparency framework (IISD, Nov. 2016).

By the conclusion of COP22, the Subsidiary Body for Implementation (SBI) had officially welcomed information reported by the GEF on the establishment of the CBIT, its programming and implementation modalities, and the pledges it received from several countries, and recommended that the COP request that the GEF continue to provide such information in its annual reports. The SBI also encouraged developing countries to submit project proposals for accessing CBIT Trust Fund resources and the GEF to 'approve the first set of CBIT projects as early as possible', subject to the availability of finances in the CBIT Trust Fund (ibid.).

At the 46th meeting of the subsidiary bodies (SB46), held in June 2017, informal consultations on modalities, procedures and guidelines for the Paris transparency framework continued, with 'growing focus' on information needed to track progress made towards goals laid out in the Parties' Nationally Determined Contributions, and an emphasis on linkages between the transparency framework and the upcoming global stocktake (IISD 2017). In these discussions, many developing countries pointed to existing challenges in tracking support received and in 'identifying information gaps' related to financial, capacity building and technology transfer needs. Divergence between developed and developing countries' stances on incorporation of differentiation into transparency guidance remained evident, with developing countries arguing for binary transparency requirements and developed countries contending that the 'nationally-determined nature' of inputs to the Paris Agreement had already accounted for necessary differentiation. While this split precluded significant progress at this meeting towards decisions on the transparency rulebook slated for adoption in 2018, some observers contended that it did represent a step forward in terms of laying out the issues at hand (ibid.).

At the conclusion of the June 2017 meeting, the Ad hoc Working Group on the Paris Agreement (APA) invited Parties to submit possible headings, subheadings, and operational details for the modalities, procedures and guidelines of the Paris transparency framework. It requested also that the Secretariat organize a roundtable to consider the Parties' submissions in advance of COP23 (ibid.).

Following the official opening of the CBIT at COP22 in November 2016, the GEF has accepted donations to the CBIT Trust Fund and worked to approve projects for CBIT financing. During COP22, 12 donors (Australia, Canada, Germany, Italy, Japan, the Netherlands, New Zealand, Sweden, Switzerland, the United Kingdom, the United States, and the Walloon Region of Belgium) issued a joint statement pledging over \$50 million to the CBIT (GEF 2017). As of April 2017, the Trust Fund had received \$43.5 million of promised contributions and additional donors had highlighted their intention to pledge to the Trust Fund 'in the near future'.

As of August 2017, the GEF has approved 11 CBIT projects, ten of which aim to strengthen the capacity of a single recipient country and one of which is a global coordination project (Table 8.1). Together, the projects amount to $11.2 million of GEF funding.

While the effectiveness of these projects remains to be seen, the CBIT is clearly beginning the process of building national capacity on collecting and reporting climate data. The projects can be considered pilot projects, but it remains to be seen whether they can and will be scaled up to regions and country groups most needing this support. And the CBIT's longer-term picture is unclear, since it was created with a single round of pledges from developed nations.

Tools, modalities, procedures and guidelines for transparency

In some ways, the only thing binding and enforceable that came out of the Paris negotiations was a system of transparency. Paris built a new 'enhanced transparency framework' to monitor, report and review information on Parties' greenhouse gas (GHG) emissions, actions taken to reduce those emissions and what they are doing to adapt to the impacts of climate change, as well as the financial, technological and capacity-building support provided and received by some Parties.

The UNFCCC in Article 12 requires that all Parties submit regular national reports, in the form of National Communications (NCs). Annex I and non-Annex I Parties have different requirements (see Van Asselt et al. 2017), and revised guidelines for Annex I Parties are currently under consideration. Parties agreed to make the National Communications submitted by Annex I Parties every four years subject to regular in-depth reviews. These reviews are organized by the UNFCCC Secretariat and are carried out by Expert Review Teams (ERTs), which are comprised of experts nominated by Parties and, at times, from intergovernmental organizations. National Communications submitted by non-Annex I Parties are not subject to review.

In addition to National Communications, all Parties need to submit regular greenhouse gas inventories, with Annex I Parties required to do so on an annual basis. These reports consist of a National Inventory Report and a Common Reporting Format, which provides the main information in table form. The Cancun Agreements specify that Annex I Parties need to submit new Biennial Reports (BRs) every two years, either independently or together with their National Communications. The Biennial Reports are subject to International Assessment and Review (IAR), a process that combines a technical expert review with a new peer-to-peer process called Multilateral Assessment (MA).

Table 8.1 'Concept approved' CBIT projects, as of August 2017

Project title	*Location*	*Agency*	*GEF grant ($)*	*Cofinancing ($)*
Strengthening capacity in the agriculture and land-use sectors for enhanced transparency in implementation and monitoring of Cambodia's NDC	Cambodia	FAO	863,242	1,731,000
Strengthening Chile's NDC Transparency Framework	Chile	UNEP	1,232,000	870,000
Strengthening capacity in the agriculture and land-use sectors for enhanced transparency in implementation and monitoring of Mongolia's NDC	Mongolia	FAO	863,242	1,160,000
Strengthening capacity in the agriculture and land-use sectors for enhanced transparency in implementation and monitoring of NDCs in Papua New Guinea	Papua New Guinea	UNEP	863,242	1,550,000
Strengthening Ghana's national capacity for transparency and ambitious climate reporting	Ghana	UNEP	1,100,000	1,310,000
Strengthening the capacity of institutions in Uganda to comply with the transparency requirements of the Paris Agreement	Uganda	CI	1,100,000	450,000
Building institutional and technical capacities to enhance transparency in the framework of the Paris Agreement	Uruguay	UNDP	1,100,000	760,000
CBIT Global Coordination Platform	Global	UNEP	1,000,000	400,000
Strengthening national institutions in Kenya to meet the transparency requirements of the Paris Agreement and sharing best practices in the East Africa Region	Kenya	CI	1,000,000	1,050,000
Capacity-building programme to implement South Africa's Climate National System	South Africa	UNEP	1,100,000	2,289,065
Costa Rica's integrated reporting and transparency system	Costa Rica	UNEP	1,000,000	3,260,000
Total financing committed as of August 2017			11,221,726	14,830,065

Cancun also introduced new obligations and processes for developing country Parties, who agreed to submit Biennial Update Reports (BURs) every two years from 2014 onwards – with the exception of Least Developed Countries (LDCs) and Small Island Developing States (SIDS), who can do so at their discretion.

These reports are subject to International Consultation and Analysis (ICA) under the SBI. The aim of the International Consultation and Analysis is to enhance transparency through a process that is to be non-confrontational and non-intrusive, and that respects national sovereignty. The process mirrors the two steps of the IAR (International Assessment and Review) that developed countries go through, by starting with an analysis of Biennial Update Reports by a team of technical experts, in consultation with a Party. Based on the experts' report, a Facilitative Sharing of Views (FSV) will take place, which can include questions and answers between Parties. The first such sessions took place during two SBI workshops in 2016, covering a total of 20 developing country Parties, including Brazil, Mexico, South Africa, and South Korea. For the purposes of the International Consultation and Analysis, LDCs and SIDS can be analysed in groups, rather than individually.

By early 2017, only 36 developing countries had submitted their first Biennial Update Reports, which were due by the end of 2014. Although reporting requirements for developing countries are less stringent than those for developed countries, this suggests that developing countries are experiencing difficulties with aspects of reporting (or that this is not a priority for them). This may be related, among others reasons, to a lack of financial resources, data, or established domestic reporting infrastructures (Ellis & Moarif 2015). In other words, reporting challenges are associated with capacity constraints.

The debate on modalities of accounting and reporting climate finance continues, and there are also issues of how to report on climate change capacity-building efforts supported internationally. Weikmans (2017) reports that there is vast differentiation in how countries are actually doing on transparency of reporting action and support received in their Biennial Update Reports. Finally, it is important to acknowledge that there are other efforts to build capacity on transparency that already exist outside of the CBIT, especially

> [the] International Partnership on Mitigation and MRV, which Germany, South Africa and South Korea have been developing since 2010 and is now used by over 70 developing, emerging and industrialised nations to exchange information and experience on climate change mitigation.
>
> (GIZ n.d.; Partnership on Transparency in the Paris Agreement n.d.)

The group was renamed the 'Partnership on Transparency in the Paris Agreement' in 2016. They describe themselves as a 'semiformal forum', which seeks to: 'foster transparency, communication, networking and trust between countries, provide capacity building and promote a mutual learning process within regions and among practitioners at a global scale, and identify and disseminate good practices examples and lessons learned'. These goals clearly overlap with the CBIT efforts, but focus mostly on mitigation through

> Partnership Meetings held on the fringes of UN climate negotiations, Capacity building activities and peer-to-peer learning through technical workshops in five regional and language groups, and through international conferences, and Knowledge products and knowledge sharing, for example through newsletters and a website.
>
> (Partnership on Transparency in the Paris Agreement n.d.)

There are a few other initiatives in the different areas of climate mitigation, adaptation, finance and technology transfer that all seek to improve information flows and learning in these areas.

Barriers in reporting climate action and directions for improvement: insights from Jamaica

For the case of Jamaica discussed in Chapters 5 and 7, we sought to explore a few questions on capacity in reporting their climate action. First, what is the level of capacity in the country to collect data and reporting to the UNFCCC on emissions and adaptation efforts? Second, what is the level of capacity for collecting data on support received (including its use, impact and estimated results)? On finance, we asked whether there was an 'aid management platform' in place for collecting and reporting project funding received from the range of foreign donors at work in the country. And, finally, we sought to understand what would be the most useful parts of capacity to build/develop in each country.

On the level of capacity in the country, to collect emissions and adaptation data, the Climate Change Division (CCD) in the Government of Jamaica – currently housed under the Ministry of Economic Growth and Job Creation – oversees the implementation of Jamaica's obligations as a Party (Government of Jamaica 2016). The Ministry is headed by the Prime Minister, but one of two Ministers without Portfolio is directly responsible for climate change, environment, land and investments (Ministry of Economic Growth and Job Creation 2017a; Office of the Prime Minister 2017a). In helping to fulfil Jamaica's UNFCCC obligations, the Climate Change Division prepares the national greenhouse gas (GHG) inventories, which form part of the country's National Communications and Biennial Update Reports (BURs) to the UNFCCC (Government of Jamaica 2016). The Division coordinates Jamaica's involvement in national, regional and international climate actions (ibid.). It is currently staffed by a Principal

Director, who is also Jamaica's National Focal Point for the UNFCCC, two Senior Technical Officers – one for mitigation and the other for adaptation – and a secretary (Ministry of Economic Growth and Job Creation 2017b; UNFCCC Secretariat 2017b; Williams-Raynor 2016a, 2016c). When the Division was established in 2013, it was housed under the then Ministry of Water, Land, Environment and Climate Change (Dekens & Price-Kelly 2016). With the change in political administration in February 2016, the Division became part of the newly formed Ministry of Economic Growth and Job Creation in March 2016 (Ministry of Economic Growth and Job Creation 2017a).

The CCD's work is supported by a Climate Change Focal Point Network and a Climate Change Advisory Board (Dekens & Price-Kelly 2016). The Focal Point Network, which was established in 2014, comprises approximately 27 representatives from various government Ministries, Agencies and Departments (ibid.; Williams-Raynor 2017). The Network regularly reports to the Climate Change Division and is 'responsible for supporting the integration of climate adaptation planning at the sectoral level' (Dekens & Price-Kelly 2016: 3).

Neither the Focal Point Network nor the Advisory Board explicitly collects data on GHG emissions. For the purposes of reporting on emissions, however, data are supplied to the CCD by both public and private sector actors, including the Ministry of Science, Energy and Technology, the Petroleum Cooperation of Jamaica and the Jamaica Public Service Company Limited (Government of Jamaica 2016). The Energy Ministry, in collaboration with the CCD and the Organización Latino Americana de Energía, developed the country's first Nationally Appropriate Mitigation Action template (ibid.). The Government submitted its First BUR to the UNFCCC in November 2016 (UNFCCC Secretariat 2017c).

Background work for the production of the emissions inventory chapter in the First BUR was supported by a US$852,000 grant from the Global Environment Facility (Government of Jamaica 2016; Global Environment Facility 2017d). This amount was supplemented by US$200,000 in cofinancing – US$90,000 from the United Nations Development Programme (UNDP), the project's Implementing Agency, and US$110,000 in in-kind support from the government (UNDP 2017) The funds were used: (1) to procure the services of national and international GHG experts; (2) to conduct stakeholder consultations for data gathering; and (3) to develop a position paper for and support the participation of Jamaican delegates in the Twentieth Conference of the Parties to the UNFCCC in Lima, Peru, in 2014 (United Nations Development Programme 2017). Using the Revised 2006 IPCC Guidelines for National Greenhouse Gas Inventories, particularly the default methodologies (Tier 1), Jamaica's Report compiled emissions estimates from agriculture, energy, waste, land use change and forestry, and industries for carbon dioxide, methane, nitrous oxide and hydrofluorocarbons for the years 2006–2012 (Intergovernmental Panel on

Climate Change 2006; Government of Jamaica 2016; Williams-Raynor 2015). At the time of this writing, the Government of Jamaica was in the process of preparing its Third National Communication to the UNFCCC.

The National Communication process is being led by a local consultant with the support of four other local consultants (Jamaica Information Service 2017; Serju 2017; United Nations Development Programme 2017c). The lead consultant is a retired civil servant who represented the Government in UNFCCC negotiations and in other international climate-related fora, including serving as Chairman of the Clean Development Mechanism Executive Board (IDEAcarbon Limited 2010). Aether Limited, a UK-based firm that provides air quality and climate change emissions inventory, forecasting and policy analysis consulting services, is also helping to compile and improve the country's national GHG emissions inventory (Aether Limited 2015, 2017b).

The Planning Institute of Jamaica (PIOJ), an Agency of the Office of the Prime Minister, manages external cooperation programmes and agreements, and, along with regional and international funding agencies, coordinates the identification and implementation of development projects and interventions, and it includes a Sustainable Development and Regional Planning Division (Planning Institute of Jamaica 2008, 2017). That division now focuses on vulnerability and impact assessments, marine and coastal management, ecosystem-based adaptation, and public awareness raising and training (UNEP 2017). The PIOJ has been engaging in accessing climate financing and coordinating climate change-related projects (Williams-Raynor 2016a). The PIOJ established Thematic Working Groups which aim to meet each quarter to

> Plan, implement, monitor and evaluate strategic priorities and actions; Track indicator progress; Identify and mobilize resources for the sector or thematic area; Promote and participate in the development of new policies, strategic initiatives and projects; Share information, knowledge and expertise; and Ensure concerted and coordinated technical support towards national development.
>
> (Planning Institute of Jamaica 2014)

These working groups provide an opportunity for inter-agency communication on international support received – each group is headed by a Permanent Secretary and comprises senior representatives of the government, private sector, civil society and international development partners (ibid.).

The Climate Change Division is responsible for reporting on Jamaica's implementation of the UNFCCC; prior to its establishment, reporting was largely led by the Meteorological Service of Jamaica (UNFCCC Secretariat 2014a, 2014b). The Ministry of Foreign Affairs and Foreign Trade also reports internationally on the use of support received but not necessarily on the impact and estimated results of the support (pers. comm., 27

November 2014). As was the case with the preparation of its First BUR, Jamaica receives bilateral support for meeting its reporting obligations under the UNFCCC from the European Union, the United States Agency for International Development, and the governments of Japan and Germany, among others (Government of Jamaica 2016). The country also receives multilateral support from the UNFCCC Secretariat, the UNDP, the United Nations Food and Agriculture Organization and the United Nations Environment Programme, for example (ibid.). The Caribbean Community Climate Change Centre and the Caribbean Development Bank also provide climate-related support to the government (ibid.). The government reports generally on the impact and results of internationally-funded climate change projects to the public by coordinating project-supported stakeholder forums. In 2014, for example, the PIOJ coordinated the Climate Investment Funds (CIF) Partnership Forum (Planning Institute of Jamaica 2015).

There is currently no 'aid management platform' in Jamaica – government ministries, agencies and departments individually collect data on the project funding they receive, though these are not usually made public. In view of the absence of an aid management platform, however, there are existing platforms that could be expanded to track aid. The National Summary Data Page (NSDP) is one such platform, which provides 'the most up-to-date economic, financial and socio-demographic statistics for all agencies on a single web page' (Jamaica Information Service 2017). JamStats is another platform; it tracks Jamaica's progress on development indicators and institutionalizes DevInfo, and is a 'database system for monitoring human development' (United Nations Children's Fund 2017; JamStats Secretariat 2012, 2017a).

In 2013, the UNDP conducted a study on the resources that would be required to establish a Climate Change Department within the government (UNDP 2013). CCD's technical staff complement has, however, remained unchanged and 'so the Division has a challenge of not being able to bring together all the roles that it should' (Williams-Raynor 2016a). As a result of this, the Division tries to take 'a collaborative approach, partnering with various agencies and individuals [in order] to get the work done' (ibid.). When the CCD's previous Principal Director demitted office in 2016, he underscored the need for increasing the staff complement of the CCD in order to boost the Division's efficiency (ibid.). This staff complement, he recommended, 'should include not only a communications/behaviour change officer whose services the CCD [had] already taken steps to secure but also a technical officer with responsibility for climate finance' (ibid.). The former Principal Director had also called for the strengthening of the Climate Change Focal Point Network to ensure that designated representatives are 'persons at a fairly senior level' who will have the authority to develop and implement necessary climate actions (ibid.). Jamaica's First BUR also highlights the need for increased financial support to meet UNFCCC reporting requirements (Government of Jamaica 2016).

There is a need to improve Jamaica's capacity to collect and analyse scientific data to support the country's climate adaptation and mitigation agenda. The report of Jamaica's greenhouse gas inventory for the years 2000–2005, which was part of its Second National Communication to the UNFCCC, submitted in November 2011, highlights a number of data gaps. These gaps relate to, among other things, data comprehensiveness and coding. In the energy sector, for example, fuel consumption data for all mining activities are not reported (Claude Davis & Associates Limited 2010). Land use and land use changes categories are based on outdated satellite imagery (Claude Davis & Associates Limited 2010). McCalla (2012) recommends improving and rationalizing the hydrometric network, using GIS to improve project planning and design, installing additional flood early warning systems and tide gauges, and procuring river and reservoir modelling software such as RiverWare® along with additional automatic weather stations.

Conclusion

The aim of this chapter has been to describe the major Capacity Building Initiative for Transparency that came out of Paris, and to more broadly overview existing capacity gaps that hinder developing nations' reporting of mitigation action, adaptation efforts, and reception of climate finance and transferred technology.

Indeed, transparency-related capacity building efforts will be a crucial factor determining whether the new Paris requirements succeed or fail, as the whole governance system is built upon mutual observation between Parties.

The CBIT's establishment shows a recognition of capacity building's importance, and the need to support nations reporting their climate actions and funding provided and received. Transparency levels will likely continue to differ sharply across the range of developing countries. The initial pledges of $53 million to the CBIT are encouraging, but there is still a lack of funding source that would make the programme sustainable over the longer term and scalable to more countries. Table 8.1 showed the first 11 projects funded 'in concept', which in fact only addressed capacity issues on reporting in about a dozen countries. The most capacity-challenged countries number over one hundred. Voluntary contributions are admirable, but a systematic, more adequate approach is needed.

So far, the CBIT has funded a variety of project types, but all have been fairly specific and therefore limited in their long-term and broader impacts on national transparency capacity. A project-based approach to employing CBIT funds is likely to improve only the transparency of individual submissions or other ad-hoc climate-related activities in developing countries, not long-term capacities for greater transparency. Instead, CBIT funds should be disbursed to promote sustainable mechanisms for continuous transparency-related capacity building within developing countries, including by supporting national institutions (such as universities), in a manner that allows for national ownership of the capacity-building efforts. In other words, capacity building for transparency should be viewed as a worthy end in and of itself, not simply a means to the end of achieving transparency in singular submissions or activities.

In one interview we conducted with Amal-Lee Amin, Director of the Climate Change division at the Inter-American Development Bank (July 2017), she described how the bank tries to hire consultants from the country or at least the region, and seeks in its funding to build the capacity in ministries to implement Nationally Determined Contributions, including on finance received. She described the difficulty with turnover in national agencies, that make capacity a difficult thing to maintain, once built. One problem, ironically, is that some of the best people with the most experience working with international agencies tend to move up to positions in those multilateral agencies, or move to the private sector. The solution is not a simple one, but clearly there needs to be a stronger pipeline of capable people coming up through the system, and countries should focus on building capacity in a deeper sense, through institution building, focus on education, and development of systems that can quickly train new people with the acquired wisdom of those leaving their posts. The final issue is one that might be quite disruptive in cases like Jamaica, where new people are hired for positions supported by external actors, and the drying up of this external support threatens the capacity that has been built there. Here, our central thesis of the book appears appropriate and practicable: universities should be the central hub of capacity building.

References

Aether Limited. (2015). News: Chris Dôre working with Jamaican Government on GHG emissions. Available at: www.aether-uk.com/News/2015/Chris-Dore-working-with-Jamaican-Government-on-GHG (accessed 19 July 2017).

Aether Limited. (2016). News: Chris Dore supports Jamaica with GHG emissions reporting. Available at: www.aether-uk.com/News/March-2016/Chris-Dore-supports-Jamaica-with-GHG-emissions-rep (accessed 19 July 2017).

Aether Limited. (2017a). About us. Available at: www.aether-uk.com/About-us (accessed 19 July 2017).

Aether Limited. (2017b). Jamaica GHG inventory review. Available at: www.aether-uk.com/Case-studies/Jamaica-GHG-inventory-review (accessed 19 July 2017).

Claude Davis & Associates Limited. (2010). *Final report: Jamaica's greenhouse gas mitigation assessment*. Toronto: Claude Davis & Associates.

Dekens, J., & Price-Kelly, H. (2016). sNAPshot: Jamaica's approach to initiating sector integration of adaptation considerations. In *Country Brief 1A*, edited by NAP Global Network. Winnipeg: NAP Global Network.

Ellis, J., & Moarif, S. (2015). Identifying and addressing gaps in the UNFCCC reporting framework. OECD/IEA Climate Change Expert Group Papers, No. 2015/07,=. Paris: OECD. Available at: http://dx.doi.org/10.1787/5jm56w6f918n-en

GEF. (2016a). New GEF fund gives boost to Paris Agreement implementation. Available at: www.thegef.org/news/new-gef-fund-gives-boost-paris-agreement-implementation

GEF. (2016b). *50th GEF council meeting, June 07–09, 2016: Programming directions for the Capacity-Building Initiative for Transparency*. Washington, DC: Global Environment Facility. Available at: www.thegef.org/gef/sites/thegef.org/files/documents/EN_GEF.C.50.06_CBIT_Programming_Directions_0.pdf (accessed 29 October 2016).

GEF. (2017a). Third National Communication (TNC) and Biennial Update Report to the UNFCCC. Available at: www.thegef.org/project/third-national-communication-tnc-and-biennial-update-report-unfccc-0 (accessed 20 July 2017).

GEF. (2017b). Progress Report on the Capacity-Building Initiative for Transparency. Available at: www.thegef.org/sites/default/files/council-meeting-documents/EN_GEF.C.52.Inf_.07_Progress_Report_on_CBIT_0.pdf

(GEF). (2017c). Progress Report on the Capacity-Building Initiative for Transparency. Available at: www.thegef.org/sites/default/files/council-meeting-documents/EN_GEF.C.52.Inf_.07_Progress_Report_on_CBIT_0.pdf

GIZ. (n.d.). International Partnership on Mitigation and MRV. Available at: www.giz.de/en/worldwide/30180.html (accessed 16 October 2017).

Government of Jamaica. (2013). *Green Paper: Climate change policy framework and action plan*. Edited by Land Ministry of Water, Environment and Climate Change. Kingston: Government of Jamaica.

Government of Jamaica. (2016). *Biennial update report of Jamaica*. Bonn: United Nations Framework Convention on Climate Change Secretariat.

IDEAcarbon Limited. (2010). Interview: Clifford Mahlung – Chairman of the CDM Executive Board. Available at: www.ideacarbon.com/media-and-events/index.htm/Interview%20Clifford%20Mahlung%20-%20Chairman%20of%20the%20CDM%20Executive%20Board (accessed 22 July 2017).

Intergovernmental Panel on Climate Change. (2006). *2006 IPCC guidelines for national greenhouse gas inventories*. Geneva: Intergovernmental Panel on Climate Change.

International Institute for Sustainable Development (IISD). (2016). Summary of the Bonn Climate Change Conference: 16–26 May 2016. Available at: http://enb.iisd.org/download/pdf/enb12676e.pdf

IISD. (2016). Summary of the Marrakech Climate Change Conference: 7–19 November 2016. Available at: http://enb.iisd.org/download/pdf/enb12689e.pdf

IISD. (2017). Summary of the Bonn Climate Change Conference: 8–18 May 2017. Available at: http://enb.iisd.org/download/pdf/enb12701e.pdf

Jamaica Information Service. (2017). Tender: Ministry of Water, Land, Environment and Climate Change. Government of Jamaica. Available at: http://jis.gov.jm/procurement/ministry-water-land-environment-climate-change/ (accessed 29 July 2017).

Jamaica Information Service. (2017). Tender: Ministry of Water, Land, Environment and Climate Change. Government of Jamaica,. Available at: http://jis.gov.jm/procurement/ministry-water-land-environment-climate-change/ (accessed 29 July 2017).

JamStats Secretariat. (2012). Overview of JamStats. Government of Jamaica. Available at: www.jamstats.gov.jm/Overview/tabid/61/Default.aspx (accessed 27 July 2017).

JamStats Secretariat. (2017a). The Secretariat. Government of Jamaica. Available at: www.jamstats.gov.jm/AboutUs/tabid/55/Default.aspx (accessed 27 July 2017).

JamStats Secretariat. (2017b). Welcome to JamStats. Government of Jamaica. Available at: www.jamstats.gov.jm/ (accessed 27 July 2017).

McCalla, W. (2012). *Review of policy, plans, legislation and regulations for climate resilience in Jamaica*. Kingston: Planning Institute of Jamaica.

Ministry of Economic Growth and Job Creation. (2017a). About us. Government of Jamaica. Available at: www.mwh.gov.jm/#!/about-us (accessed 17 July 2017).

Ministry of Economic Growth and Job Creation. (2017b). Divisions and Branches. Government of Jamaica. Available at: www.mwh.gov.jm/#!/divisions (accessed 17 July 2017).

Office of the Prime Minister. (2017a). Cabinet Minister: The Honourable Daryl Vaz, MP. Government of Jamaica. Available at: http://opm.gov.jm/cabinet_ministers/daryl-vaz/ (accessed 17 July 2017).

Office of the Prime Minister. (2017b). Ministry of Economic Growth and Job Creation. Government of Jamaica. Available at: http://opm.gov.jm/portfolios/ministry-of-economic-growth-and-job-creation/ (accessed 17 July 2017).

Office of the Prime Minister. (2017c). Planning Institute of Jamaica (PIOJ). Government of Jamaica. Available at: http://opm.gov.jm/opm_agency/planning-institute-of-jamaica-pioj/ (accessed 17 July 2017).

Partnership on Transparency in the Paris Agreement. (n.d.). About the Partnership. Available at: www.transparency-partnership.net/about-partnership (accessed 16 October 2017).

Planning Institute of Jamaica. 2008. Main functions and responsibilities. Government of Jamaica. Available at: www.pioj.gov.jm/AboutUs/Functions/tabid/108/Default.aspx (accessed 17 July 2017).

Planning Institute of Jamaica. (2014). Thematic Working Groups—TWGs—Implementing, Monitoring & Evaluating Vision 2030 Jamaica. Available at: www.vision2030.gov.jm/Portals/0/Thematic_Group/TWG%20BLURB.pdf (accessed 19 July 2017).

Planning Institute of Jamaica. (2015). *Annual report 2014*. Kingston: Planning Institute of Jamaica.

Planning Institute of Jamaica. (2017). Director General and Board of Directors. Government of Jamaica. Available at: www.pioj.gov.jm/AboutUs/Board/tabid/110/Default.aspx (accessed 17 July 2017).

Serju, C. (2017). Jamaica steps up climate change resilience efforts. *The Jamaica Gleaner*. Available at: http://jamaica-gleaner.com/article/news/20170413/jamaica-steps-climate-change-resilience-efforts (accessed 16 July 2017).

UNDP. (2017c). Support to the Third National Communication the United Nations Framework Convention on Climate Change. United Nations Development Programme. Available at: www.jm.undp.org/content/jamaica/en/home/operations/projects/climate-and-disaster-resilience/support-to-the-third-national-communication-the-united-nations-f.html (accessed 16 July 2017).

UNFCCC Secretariat. (2014a). Full library record: Jamaica's first national communication to the United Nations Framework Convention on Climate Change

(UNFCCC). Available at: http://unfccc.int/essential_background/library/items/3599.php?rec=j&priref=2751#beg (accessed 20 July 2017).
UNFCCC Secretariat. (2014b). Full library record: The second national communication of Jamaica to the United Nations Framework Convention on Climate Change. Available at: http://unfccc.int/essential_background/library/items/3599.php?rec=j&priref=2751#beg (accessed 20 July 2017).
UNFCCC Secretariat. (2017a). Biennial Update Reports. Available at: http://unfccc.int/national_reports/non-annex_i_parties/biennial_update_reports/items/9186.php (accessed 20 July 2017).
UNFCCC Secretariat. (2017b). Jamaica: national focal points. Available at: http://unfccc.int/tools_xml/country_JM.html (accessed 20 July 2017).
UNFCCC Secretariat. (2017c). Submitted Biennial Update Reports (BURs) from Non-Annex I Parties. Available at: http://unfccc.int/national_reports/non-annex_i_natcom/reporting_on_climate_change/items/8722.php (accessed 20 July 2017).
United Nations Children's Fund. (2017). About DevInfo. Available at: www.devinfo.org/libraries/aspx/AboutDevInfo.aspx?T=ADI&PN=diorg/di_about.html (accessed 30 July 2017).
United Nations Development Programme. (2013). *Establishment of a Climate Change Department in Jamaica: UNDP's Analytical Report to the Ministry of Water, Land, Environment and Climate Change, Government of Jamaica*. Kingston: United Nations Development Programme.
United Nations Development Programme. (2015a). *2014 Annual Status Report: Global Project on Capacity Development for Aid Effectiveness*. New York: United Nations Development Programme.
United Nations Development Programme. (2015b). *Project document: Support to effective implementation project*. Kingston: United Nations Development Programme.
United Nations Development Programme. (2017). Support to the Third National Communication the United Nations Framework Convention on Climate Change. Available at: www.jm.undp.org/content/jamaica/en/home/operations/projects/climate-and-disaster-resilience/support-to-the-third-national-communication-the-united-nations-f.html (accessed 16 July 2017).
United Nations Environment Programme. (2017). Planning Institute of Jamaica (PIOJ). Available at: www.cambioclimatico-regatta.org/index.php/en/key-institutions/item/planning-institute-of-jamaica-pioj (accessed 12 July 2017).
Van Asselt, H., Weikmans, R., Roberts, J. T., & Abeysinghe, A. (2016). Transparency of action and support under the Paris Agreement. Available at: https://ssrn.com/abstract=2859151.
Van Asselt, H., Weikmans, R., & Roberts, J. T. (2017). *Pocket guide to transparency under the UNFCCC*. Oxford: European Capacity Building Initiative.
Weikmans, R. (2017). Transparency from the other side: A review of the first Biennial Update Reports. In *AdaptationWatch 2017 Report: Implementing adaptation: building accountability, capacity, and finance*. Online at AdaptationWatch.org.
Williams-Raynor, P. (2015). Jamaica takes stock of greenhouse gas emissions. *The Jamaica Gleaner*. Available at: http://jamaica-gleaner.com/gleaner/20150123/news/news5.html (accessed 21 July 2017).
Williams-Raynor, P. (2016a). Albert Daley bids farewell to Climate Change Division. *The Jamaica Gleaner*. Available at: http://jamaica-gleaner.com/article/

news/20160818/albert-daley-bids-farewell-climate-change-division (accessed 16 July 2017).

Williams-Raynor, P. (2016b). New climate change board holds first meeting. *The Jamaica Gleaner*. Available at: http://jamaica-gleaner.com/article/news/20161215/new-climate-change-board-holds-first-meeting (accessed 15 July 2017).

Williams-Raynor, P. (2016c). Woman in charge – Climate Change Division gets female head. *The Jamaica Gleaner*. Available at: http://jamaica-gleaner.com/article/news/20160908/woman-charge-climate-change-division-gets-female-head (accessed 16 July 2017).

Williams-Raynor, P. (2017). Jamaica to strengthen climate change focal point network. *The Jamaica Gleaner*. Available at: http://jamaica-gleaner.com/article/news/20170601/jamaica-strengthen-climate-change-focal-point-network (accessed 16 July 2017).

9 Conclusion

Implementing the Paris framework on capacity building

Introduction

We have started to identify and describe the problems underlying the capacity-building efforts that development agencies have carried out over the last several decades. These lingering problems clearly necessitate the search for a fresh approach to the whole objective of building capacity for efforts to address climate change. Such a new approach is exactly what our book attempts to devise: a way to build long-term capacity using developing countries' most sustainable institutions – universities.

This book's introductory chapter provides the context, explaining capacity building as a framework for international environmental cooperation and, in the process, challenging whether the enormous amounts of time and money that have been invested in capacity building have yielded real, lasting results. It raises important questions that the rest of the book further explores: After decades of capacity-building efforts, what is the end result? What capacities have been built, to what extent, and at what levels – individual, institutional, or systemic? Which actors have led the process, and has the process been largely demand- or supply-driven? Has any sustainable system to continuously build capacity been left in place? What are the most significant gaps in the process? The introductory chapter contends that the establishment of the Paris Committee on Capacity Building at COP21 signalled that the time has come to think of and implement a fresh approach to capacity building, consisting of a global strategy that can overcome existing inadequacies and inefficiencies to achieve long-term, sustainable capacity improvements.

Chapter 2 traces the evolution of the concept of capacity building starting in the 1950s, then zeroes in on the agenda of capacity building for climate change, beginning in 1992 with the adoption of the UNFCCC and stretching to efforts to implement the Paris Agreement in 2017. This chapter overviews the evolution of institutions, mechanisms, processes, and funding dedicated to promoting the agenda of capacity building under the UNFCCC. Since decision-making through negotiations between 195 disparate Parties to the UNFCCC is very much a political process, this history also teases out the politics that contributed to adoption of the main elements of capacity building under the UNFCCC. These politics are created by the often-conflicting interests of development agencies, development consultants, NGOs, universities and think-tanks, donor countries, and recipient countries in having capacity building happen in certain ways.

Lessons from capacity building in other regimes

Following from the second chapter's historical review of capacity building efforts under the UNFCCC, our third chapter traces the development of capacity-building initiatives in a few other multilateral development and environmental regimes. Examining the experiences of various multilateral regimes and development approaches with capacity building and considering the similarities and differences among them proved a useful exercise for devising a new approach to capacity building under the Paris Agreement. The areas of capacity building considered are: trade capacity building (TCB), capacity building under the Regional Seas Programme (RSP), capacity building to integrate human rights into development, capacity building for disaster risk management (DRM) and capacity building for phasingout ozone-depleting substances under the Montreal Protocol. Our review shows that there are many commonalities and little variation between approaches and tools used in capacity building across these regimes. Significant patterns observed across regimes are summarized below:

- *Effective capacity building focuses on institutional development and strengthening.* Where institutional development and strengthening were not a central goal, the regime did not function effectively. An example is the Mediterranean Regional Seas Programme's capacity-building initiative, which did not endeavour to strengthen states' institutions and therefore was made less effective by the fact that the public sector capacity across the various states fell short of a critical mass. Capacity-building efforts were hindered by the wide diversity of state regimes and value systems and uneven levels of development and political stability across states. On the other hand, the Regional Seas Programme capacity-building efforts in the Baltic Sea were successful because the state regimes and value systems were already sufficiently similar. Institution-focused capacity building has also proven successful under the Montreal Protocol, especially with its use of detailed programme and implementation plans at both national and international levels.
- *Education, training and research aimed at human resource development and improving professional competence on a sustainable basis are also key to building capacity.* This was evident across the regimes we examined, although each placed varying emphasis on formal

education versus training. The regimes on trade, human rights and disaster risk management focused on formal education at different levels, including development of research and analytical capacities, while the Montreal Protocol regime placed greater focus on training and awareness-raising than on education.

- *Strong financial support for capacity-building efforts led to greater compliance across regimes.* Robust funding for capacity building under the trade and ozone regimes, which donor countries made available out of their direct interest in supporting developing country regime compliance, contributed greatly to developing countries' capacities to abide by the regime provisions. This was also true to a more limited extent in the case of the human rights regime, as improved human rights in developing countries broadly enhances aid effectiveness and is therefore also in the interest of developed countries providing aid.
- *National ownership of capacity-building efforts is key to sustainable progress.* Attempts to build developing countries' capacities to pursue initiatives in which they have direct interests, such as phasing out ozone-depleting substances, are more likely to be enthusiastically adopted by recipient governments, to be seen as nationally-driven, and, therefore, to prove successful. If aid projects are viewed as donor-driven at the outset, perceptions of country ownership typically do not increase notably over time, and this hinders capacity-building progress.
- *Web-based tools have contributed to successful capacity building, especially under human rights and disaster risk management regimes.* Mexico recently developed a web-based tool to educate citizens about human rights, which some other countries are now replicating. In addition, the UN Office for Disaster Risk Reduction created an online tool to assist countries in monitoring their progress and setbacks in implementing the recommendations of the Hyogo Framework for Action, which aims to build the resilience of nations and communities to disasters (UNISDR). This web-based tool has built governments' capacities to document their implementation achievements and, in the process, has allowed the accumulation of a comprehensive body of information on global disaster risk management progress.
- *Capacity substitution by external experts and consultants does not build lasting in-house capacity in recipient countries.* This fact was most manifest in the case of the disaster risk management regime, as external experts' relatively large role inhibited in-country capacity utilization and further development once the experts had departed. Many developing countries like Bangladesh have developed a fair amount of managerial and technical expertise in disaster risk management because of their vast experience facing natural disasters, and capacity-development efforts that build upon this expertise would be far more effective than those endeavouring to replace it with outside knowledge.

While a dominant role for external experts has been seen to inhibit capacity-building progress across regimes, networking, partnerships, and sharing of experiences have consistently made capacity-building efforts more successful.

Lessons learned for future climate-related capacity building

The primary lesson to take away here is that *sustainable* capacity support, at both the international and the national level, is key to successful capacity building. Sustainable support focuses on building and strengthening institutions, draws upon local expertise instead of relying upon external knowledge, and is backed by robust funding.

Second, shared interests in the larger objectives of capacity-building efforts (e.g. mitigating climate change by phasing out ozone-depleting substances) helps cement trust between donors and recipients and thereby enhances the effectiveness of capacity building. Therefore, an expanded understanding of other nations' interests among developed and developing countries alike could support more successful capacity building. Increased trust could encourage developed countries to allow greater ownership of capacity-building efforts by recipient countries, and to engage in networking, partnerships, and sharing of experiences instead of undertaking top-down, donor-driven approaches to capacity building. Allowing programmes to be designed and implemented through participatory, transparent processes has proven time and again to increase the effectiveness of capacity-building efforts.

Finally, our review of capacity-building efforts indicated that education, training and awareness-raising efforts aimed at human resource development are key to sustaining capacity building over the long term. In attempts to address climate change via capacity building, human resource development should be the central focus at both the national and the global level, and should be pursued by means of formal and informal initiatives alike. Therefore, climate change education must be integrated into national curricula around the world, at every level of education.

Synthesizing the framework

Chapter 4 endeavours to lay out a capacity-building framework up to the task of addressing the challenge of climate change. This framing chapter begins by identifying the types of capacity building needed to enable the world's most vulnerable countries to effectively address climate change and respond to its impacts. It explains what capacity building is and what types exist, encompassing various dimensions, levels, principles and indicators. We find that the concept of capacity building has historically been constructed vaguely and capaciously, such that many possible meanings coexist under the single heading of 'capacity building'. Therefore, we move to sketch out possible parameters of capacity building, using a foundational understanding that ownership of capacity-building efforts by developing countries and donor-recipient partnerships are crucial elements of successful capacity-building efforts.

As we have seen, the same problems – short-term planning, project-based funding, overuse of consultants, donor-driven processes and little feeling of ownership or commitment to the cause among recipient countries – that have stymied capacity-building efforts to date continue to hinder the process. In acknowledgement of these pitfalls, donor countries have initiated limited reform processes, but these have largely proven half-hearted and cosmetic. Simultaneously, academics have begun to recognize the enduring nature of these problems and coalesce around new ideas for effective capacity building. Literature published over the last decade on this topic focuses largely on the solution of establishing a partnership between funders and recipients. Research on outcomes of this approach is not yet available, but we believe it has enormous potential and see great value in exploring it conceptually for the time being.

Creating a genuine donor-recipient partnership is challenging from the outset of any capacity-building initiative, as funding and knowledge transfer from donor to recipient invariably create an uneven power dynamic. To address this problem, funding for climate change science and policy education must be increased domestically in recipient countries, while developed countries' provision of funding for capacity building must no longer be viewed as a voluntary, charitable donation, but as fulfilling a responsibility to contribute to the global good of mitigation and protect vulnerable countries and populations from the impacts of the climate crisis that developed countries very largely caused.

This funding should take the form of untied budget support for education and long-term, continuous capacity building, and transparency systems and other safeguards against misuse of funds should be built and enforced. Mobilization of sufficient international public finance for capacity building and delivery of this funding through agreed mechanisms are likely to contribute substantially to recipient country ownership and real partnership-building.

Finally, vulnerable countries have rich indigenous knowledge of effective adaptation actions, based on age-old experiential learning. This knowledge should contribute to mutual learning in attempts to build capacity to address climate change, and university partnerships across the world can spearhead efforts to integrate local knowledge with development agencies' expertise.

Having discussed genuine partnership as the cornerstone of successful capacity-building initiatives, we next embark on a review of capacity-building efforts in three case study countries, considering what areas of capacity-building work have proven successful and what capacity gaps still exist. Our selection of Bangladesh, Uganda and Jamaica covers three different geographical configurations (coastal nation, landlocked nation, and island nation) on three different continents. We begin by presenting country profiles, including overviews of basic socio-economic parameters and climate change impacts. While the socio-economic profiles of these three countries differ greatly, all three have agricultural sectors, water systems and populations highly vulnerable to climate change. Uganda suffers most from floods, droughts and increasing variability of seasonal rainfall. Bangladesh is, similarly, at great risk from floods and erratic rainfalls, but is also afflicted with intensifying cyclones and storm surges. Like Bangladesh, Jamaica is extremely vulnerable to hurricanes, storm surges and depressions.

We identify both similarities and differences between the policy-institutional frameworks of our three case study countries. All three have developed elaborate policies and plans to cope with climate change, and Uganda and Bangladesh have created a few distinct institutional mechanisms. Uganda has established a national climate change committee and advisory group, which consults many stakeholders, including civil society representatives, in constructing climate change policy. Uganda has also developed a GHG Inventory Management System and mainstreamed guidelines for climate-dedicated budgeting down to the district level. Bangladesh has developed a Climate Fiscal Framework, but it is not yet operationalized. It has also introduced a climate change budget line in its budgetary process and established a climate change trust, which includes civil society and NGO representatives. Bangladesh also differs from Uganda and Jamaica by its vibrant NGO activity in the development and environment sectors, as well as its regular inclusion of non-governmental experts in UNFCCC negotiations. Instead of focusing on developing national-level institutions and NGO networks like Uganda and Bangladesh, Jamaica's approach to confronting climate change largely centres on its regional network with other Caribbean nations, which have similar geographical vulnerabilities to climate change and capacities to respond to it.

Bangladesh and Uganda depend more on bilateral than multilateral sources of funding, getting most of their international support from aid agencies such as DFID, USAID and DANIDA. On the other hand, Jamaica receives the most funding for its regional projects from multilateral institutions, such as the World Bank, CARICOM, and the Inter-American Development Bank.

In terms of starting points for climate-related capacity building, Bangladesh has a distinct advantage in the area of disaster management and rehabilitation because of its historical experience with high vulnerability to climate disasters. This factor has made it a leader of adaptation efforts among developing countries. However, as is also the case in Uganda, disaster risk management and climate change efforts are undertaken by two separate ministries. This institutional arrangement stands in the way of coordinating mitigation and adaptation efforts and capturing synergies in both countries. Both Bangladesh and Uganda also suffer from lack of funds, expertise in technical areas and weaknesses within institutions.

After examining each of our three country case studies – capacity-building recipients – we proceed to review the donors' side of the capacity-building equation, delving into bilateral and multilateral development agencies' project profiles. Our main finding was that, predictably, capacity-building priorities lined up with the broad focus of each agency. For example, UNDP capacity-building projects tended to focus on the intersection of development and environment, and on how climate change exacerbates economic strife. Similarly, as UNEP focuses on ecosystem issues, its projects centre on how climate change affects ecosystems, fisheries and agriculture. The African Development Bank concentrates its efforts on bringing Africa into an inclusive economic growth framework, and works from a foundational understanding that the African continent is home to the world's nations least responsible for and most vulnerable to climate change. The Inter-American Development Bank focuses on cross-cutting issues like gender equality and diversity, climate change and environmental sustainability, and institutional capacity. On climate, it seeks to provide guidance in dialogues between governments, civil society and the private sector concerning policy agendas.

We find substantial variation across bilateral agencies' work, as their focuses are determined by their donor nation's specific interests and agenda. For example, the Canadian bilateral, CIDA, concentrates on women and gender issues, even when undertaking climate change and capacity-building efforts in developing countries. The UK aid agency, DFID, has always been focused on poverty reduction, and this has shaped their climate and capacity-building work. NORAD, Norway's development agency, is extraordinary in its focus on developing education and research capacity: all its relevant projects were university-based, focused on building staff capacity by creating and expanding Master's and PhD programmes on climate change research and management. 'Strengthening education and research capacity in climate change and natural resource management in universities,' as NORAD phrased its objective for a project on fisheries in Sri Lanka and Vietnam, is precisely the approach to capacity building we have advocated in Chapter 7 (NORAD 2013).

Across multilateral and bilateral agencies, we observed that attention to capacity building as an aspect of climate change management is highly uneven. Furthermore, we find that comparison of project work across agencies is seriously hindered by differences in the language agencies employ to describe their work. For example, some use the term 'adaptive capacity' while others use 'resilience' to refer to building capacity to withstand climate impacts. Finally, we remarked upon some areas that are undeveloped across agencies' portfolios. The most noteworthy is education and training in climate change science and management, which was not a focus of any agency besides NORAD.

Because knowledge transfer is more enduring if the recipient country is empowered to educate its citizenry about climate change science and management, expanding and strengthening regional and local knowledge systems and creating lasting training programmes should be a widespread priority among development agencies. Agencies should consider building the 'soft infrastructure' of educational capacity just as seriously and extensively as they consider building sea walls, planting drought-resistant crops, or supporting national communications.

Chapter 6 also discusses huge variations across and even within agencies in disclosure and transparency of their efforts at the project level. Some agencies provide detailed project documents consistently, while others give only one letter of support. Some agencies have only summaries available on the web, while others post just titles. Our review covers several examples illustrative of this wide variation: the GEF is highly transparent and provides detailed documentation for its projects, in line with other World Bank-administered programmes; Swedish SIDA projects are thin on accessible documentation; NORAD has detailed information available, including a listing of contact information for key individuals and links to news coverage of projects, but lacks detailed project descriptions; the IDB provides excellent documentation, including information on allocation of funds, mapping, project documents, and supporting letters from host governments; and the Adaptation Fund is consistent in providing project documents, and also usefully provides a clear breakdown of budget components and objectives. World Bank-administered agencies (GEF, Adaptation Fund, LDCF) had among the best transparency of all agencies examined.

The lessons learned from our country case studies and review of agency practices form the foundation on which Chapter 7 establishes our vision of universities as the central hub of capacity building under the Paris Agreement. The rationale for universities as the central hub is that universities are the most sustainable institutions in developing countries, and that empowering universities to educate students on climate change science and management can build a country's capacity to confront climate change continuously and over the long term. An explanation of our university-centred vision is followed by a presentation of our findings from fieldwork on six universities, located in Bangladesh, Uganda and Jamaica, including their existing strengths and weaknesses, their collaborations with external partners, and other factors relevant to their potential to serve as capacity-building hubs. The chapter ends by elaborating a roadmap of what can be done to make universities the core of a new global capacity-building initiative in terms of building infrastructures, gathering funding, and developing programmes, global partnerships, and networks.

The aim of Chapter 8 is to describe the new Capacity Building Initiative for Transparency, a major development that was established by the decision text associated with the Paris Agreement, and to overview existing capacity gaps that hinder developing nations' reporting of mitigation action, adaptation efforts, and reception of climate finance and transferred technology. Transparency-related capacity-building efforts will be a crucial determinant of whether the Paris Agreement succeeds or fails, as the whole governance system is built upon Parties reporting their own progress. Because capacities for transparency are currently sharply different across developing countries, and because mutual transparency and real accountability will prove critical to implement the Agreement, success in building developing countries' reporting capacities is essential. The initial pledges of $53 million to the CBIT at the Paris COP and just afterwards are certainly encouraging, but there remains a lack of reliable, sustainable funding that will ensure the programme functions over the long term and can scale to support capacity building in a sufficient number of countries. Table 8.1 shows the first set of projects the CBIT is funding, which in fact only address reporting-related capacity issues in about a dozen countries. The most capacity-challenged countries number over one hundred, so the CBIT's current activities can only be seen as a series of pilot projects. Voluntary contributions are admirable, but a more systematic approach to supporting the CBIT is needed.

Furthermore, a project-based approach to employing CBIT funds is likely to improve only the transparency of individual submissions or other ad-hoc climate-related activities in developing countries, not long-term capacities for greater transparency. Therefore, we believe CBIT funds should be distributed to promote sustainable mechanisms for continuous transparency-related capacity building within developing countries, including by supporting national institutions (such as national monitoring systems and universities), in a manner that allows for national ownership of the capacity-building and accountability efforts. In other words, capacity building for transparency should be viewed as a worthy end in and of itself, not simply as a means to the end of achieving transparency in singular submissions or activities.

Institutions and funding for implementing the framework

Our penultimate section deals with the means of implementing the framework for capacity building that we propose. One element of our proposal in serious need of greater attention and additional funding is retention of built capacity in local institutions. Ironically, some of the best people working on climate change within developing country governments tend to move, after a time, to positions at multilateral agencies or within the private sector. In addition, external funders often initially finance the hiring of highly competent people into in-country government positions, an approach that renders enhanced capacity fleeting, as the hiree will depart if and when grants dry up. Therefore, there is a clear need to create and support pipelines of qualified, well-educated individuals to staff developing countries' government agencies and conduct climate-related research. This will require extensive investment in in-country climate education, as well as exploration of means to make jobs focused on national climate change response more attractive. This is where our central thesis of the book, that universities should be the central focus of capacity-building efforts, emerges as an excellent solution. To pursue this approach, universities will likely need some endowment or general budget funds to 'back-stop' grants from foreign or national contributors, so that key staffers will be able to stay on, even through lean times.

Universities of all stripes are well equipped to become the focus of capacity-building efforts and to take on the crucial role of supplying national capacity. Older universities already have elaborate infrastructures in place and have proven themselves highly sustainable, while newer universities sometimes boast better facilities, as well as greater agility in taking up new challenges. Many universities, old and new, already offer stand-alone degree programmes either in environmental studies and sciences or climate change specifically, often with applied research and policy components. Various institutes and centres dedicated to climate research in developing countries are also involved in many partnership projects with regional and western universities, often funded by bilateral or multilateral agencies. Such partnership programmes may include student and faculty exchanges between the Global North and the Global South.

Expanding upon existing climate education programmes in the Global South, as well as partnership programmes linking universities in the Global North and the Global South, will be crucial aspects of university-focused capacity-building efforts. Flows of students and faculty between the Global North and the Global South will allow invaluable transfer of knowledge, expand southern students' and professors' access to the latest peer-reviewed literature, and expose all scholars involved to ways of thought in new parts of the world.

Existing weaknesses in southern universities – including lack of access to the latest peer-reviewed knowledge, and resulting limitations on abilities to conduct and publish cutting-edge research – are reinforced by lack of national-level funding for research, including construction of research labs and climate observation systems. In addition, in some cases, language efficiency is inadequate for advanced research and publication in prestigious journals, or even for effective communications with foreign counterparts. Therefore, university-focused capacity-building efforts must include funding dedicated to building local research capacities, supporting North-South partnerships, and improving language efficiency where necessary.

A way forward

At COP22 in November 2016, Parties to the UNFCCC adopted the Terms of Reference for the PCCB, which included a responsibility to 'address gaps and needs, both current and emerging' in the area of capacity building. Since this development, the Paris Committee on Capacity Building held its first meeting at the UNFCCC intersessional in May 2017, where its members decided to make a rolling workplan for the PCCB available for public inputs and to recommend to that COP that it invite Parties and other relevant institutions to provide appropriate support and resources for the implementation of the workplan.

As we have repeatedly emphasized, the most significant lingering gaps in capacity building are climate change education, training and public awareness. Therefore, increasing financing for climate-related education and training across the world will prove a crucial aspect of effective global action on climate change. In highly vulnerable developing countries, such initiatives should focus on enhancing higher education, as development agencies have historically focused their investments on primary education in poor countries (Psacharopoulos & Patrinos 2002). Furthermore, overall funding for capacity building is currently insufficient to support an effective global response to climate change. While financing for capacity building is extremely difficult to quantify, as capacity building is a vaguely defined, cross-cutting concept that is relevant in a great many aid projects, it is clear that funding remains poor as well as poorly understood (Chen & He 2013; Nakhooda, 2015; UNFCCC 2016). Funding for capacity building must be greatly enhanced, and should be viewed not as charitable donation, but as a fulfilment of developed countries' responsibilities to help protect poor and vulnerable peoples from the climate crisis that wealthy countries very largely created.

We have suggested that a different kind of external capacity-building funding can leverage necessary domestic funding. Khan, lead author of this book, has argued elsewhere that while greenhouse gas emissions are regarded as a negative externality, a 'global public bad,' climate change *impacts* resulting from emissions and undersupply of mitigation should be regarded as a global public bad as well (Khan 2014). Logically, together with mitigation, adaptation should be considered as a global public good. In an age of global commons problems, the conventional conception of public good (Samuelson 1954) should have an expanded interpretation. Therefore, funding for provision of environmental global public goods should be externally-sourced, beyond ODA, by means of taxing the global public bads (carbon emissions). This is the most fundamental principle of neoclassical economics, i.e. internalization of externalities to correct for market distortions. As a logical follow-up, the nature and sources of funding to build capacity to address climate change should be changed. Capacity building finance should no longer be regarded as a one-way flow of voluntary donor contributions, but a means of catering to a global good of capacity building to protect vulnerable countries and communities. As Archibugli and Filipetti (2015) argue, the normative implication of the global public goods analysis in the case of knowledge requires greater public investment and international cooperation.

Further, on top of the emerging consensus that capacity building must be financed sustainably and out of obligation, not on an ad-hoc basis or out of altruism, there is growing emphasis in academia and international negotiations alike on the importance of recipient ownership of capacity-building efforts and greater recipient control over funds provided. As discussed earlier, funding flowing from industrial countries to vulnerable countries establishes a dynamic of donor domination and promotes recipient country accountability upwards, to donors, over accountability downwards, to vulnerable communities. Possible solutions to this issue include efforts to build genuine donor-recipient partnerships, as well as use of international institutions to collect and distribute funding for capacity building and thereby limit direct control of resources by donor governments.

In response to the Paris Agreement's reaffirmation that capacity building and climate education are crucial to effective action on climate change, a number of universities around the world have begun to seriously consider and convene to discuss their roles in supporting capacity building. One significant early example of such collaboration was the 2016 founding of the Least Developed Countries Universities Consortium on Climate Change (LUCCC), a group of ten universities from the Global South dedicated to enhancing knowledge on climate change through capacity building. A second university collaboration, the University Network on Climate Change (UNCC), is just getting started, but is envisioned to be an information-sharing, mutual support network that will build climate-related capacity in both northern and southern universities. With these two new nascent networks, there is significant opportunity to expand mutual support and improve accountability if funding agencies will take up the challenge put forward by this book. The Paris Committee on Capacity Building could oversee this shift from the old model of ad-hoc, project-based capacity building to the longer-term approach of building new cohorts of climate-educated graduates through the training and research programmes we propose. There is much at stake, and much to be gained from a new approach, and the massively complex and unrelenting problem of climate change requires we think boldly and creatively towards the long term.

References

Archibugi, D., & Filippetti, A. (2015). Knowledge as global public good In D. Archibugi, & A. Filippetti (eds), *Handbook of global science, technology and innovation*. Oxford: Wiley Blackwell.

Chen, Z. & He, J. (2013). Foreign aid for climate change related capacity building. WIDER Working Paper. No. 2013/046, April.

Khan, M. R. (2014). *Toward a binding climate change adaptation regime: A proposed framework*. London: Routledge.

Nakhooda, S. (2015). Capacity building activities in developing countries. Workshop on potential ways to enhance capacity building activities. Bonn, 17 October.

NORAD. (2013). Regional capacity building for sustainable natural resource management and agricultural productivity under changing climate. Available at: www.norad.no/en/front/funding/norhed/projects/regional-capacity-building-for-sustainable-natural-resource-management-and-agricultural-productivity-under-changing-climate/

Psacharopoulos, G.. & Patrinos, H. A (2002). Returns to investment in education: a further update. Policy Research Working Paper No. 2881. Washington, DC: World Bank.

Samuelson, P. A. (1954). The pure theory of public expenditure. *The Review of Economics and Statistics*, *36*(4), 387–389.

UNFCCC. (2016). Decision-/CP.22. Third comprehensive review of the implementation of the framework for capacity-building in developing countries under the Kyoto Protocol. Available at: http://unfccc.int/files/meetings/marrakech_nov_2016/application/pdf/cmp12_i10_3rd_comprehensive_review_for_kp_cb.pdf

UNFCCC. (2017). First meeting of the Paris Committee on Capacity-building Bonn, Germany, 11–13 May 2017. Paris Committee on Capacity-building. PCCB/2017/1/10 16 June 2017.

Index

Page numbers in **bold** denote tables and boxes.